Smart Internal Stimulus-Responsive Nanocarriers for Drug and Gene Delivery

Smart Internal Stimulus-Responsive Nanocarriers for Drug and Gene Delivery

Mahdi Karimi[1,*], **Parham Sahandi Zangabad**[2,3,*], **Amir Ghasemi**[2,3,*] **and Michael R Hamblin**[4,5,6]

[1]*Department of Medical Nanotechnology, Faculty of Advanced Technologies in Medicine, Iran University of Medical Sciences, Tehran, Iran*
[2]*Department of Materials Science and Engineering, Sharif University of Technology, PO Box 11365-9466, 14588 Tehran, Iran*
[3]*Advanced Nanobiotechnology & Nanomedicine research group (ANNRG), Faculty of Advanced Technologies in Medicine, Iran University of Medical Sciences, Tehran, Iran*
[4]*Wellman Center for Photomedicine, Massachusetts General Hospital, Boston, MA 02114, USA*
[5]*Department of Dermatology, Harvard Medical School, Boston, MA 02115, USA*
[6]*Harvard MIT Division of Health Science and Technology, Cambridge, MA 02139, USA*

*These three authors contributed equally to this book.

Morgan & Claypool Publishers

Rights & Permissions
To obtain permission to re-use copyrighted material from Morgan & Claypool Publishers, please contact info@morganclaypool.com.

ISBN   978-1-6817-4257-1 (ebook)
ISBN   978-1-6817-4256-4 (print)
ISBN   978-1-6817-4259-5 (mobi)

DOI   10.1088/978-1-6817-4257-1

Version: 20151101

IOP Concise Physics
ISSN 2053-2571 (online)
ISSN 2054-7307 (print)

A Morgan & Claypool publication as part of IOP Concise Physics
Published by Morgan & Claypool Publishers, 40 Oak Drive, San Rafael, CA, 94903, USA

IOP Publishing, Temple Circus, Temple Way, Bristol BS1 6HG, UK

*Dedicated to our beloved parents, and wives
for their sacrifice through the years.*

Contents

Preface

In recent decades, innovative breakthroughs have emerged in the broad and flourishing field of nanobiotechnology. This arena of technology and its particular branch, nanomedicine, have made a significant impact on numerous fields of science and technology including materials science, biotechnology, and biomedicine. On the other hand, design of smart systems possessing controllable behavior with accurate feedbacks to different stimulations has focused the concentration of various researches in nanbiotechnology, nanomedicine, and the associated field of drug delivery systems (DDSs). Hence, innovative smart stimulus-responsive drug delivery systems have recently attracted the interest of multifarious research and studies.

This matching pair of E-books weaves together many of the strands that make up the emerging field of modern nanomedicine. Drug-delivery, controlled-release, gene therapy, nanocarriers and smart intelligent nanosystems are highly relevant to the design of stimulus-responsive drug and gene delivery systems.

Much of the motivation for the development of this field has come from an appreciation of the drawbacks of traditional cancer chemotherapy. Many of the approved drugs, which are actually quite good at killing cancer cells, are also highly toxic to normal cells. This unfortunate truth explains the high (almost universal) incidence of side-effects in cancer chemotherapy, which can rapidly become intolerable to patients and even life-threatening. Moreover, many of the drugs used in cancer chemotherapy are highly insoluble in biological media and have sub-optimal pharmacokinetics and biodistribution. A range of nanocarriers and nanovehicles has been designed to solubilize these drugs, and allow them to be transported intact in the bloodstream (after intravenous injection) until they reach their intended tumor target. But how are these nanocarriers meant to know when their target has been reached? The pressing need to find an answer to this question has been the driving force for the creation of an impressive range of smart or stimulus-responsive nanocarriers, which have been engineered at the molecular level to respond to a physical, chemical, or biological stimulus that is present at, is overexpressed at, or can be externally applied at the tumor site. It is noteworthy that considering the high potential of smart stimulus-responsive drug/gene delivery systems, they are increasingly being applied in diagnosis and therapy of other formidable disorders, infections, inflammations and diseases such as Alzheimer's, cardiovascular diseases, diabetes, etc, and are prompting newfound and efficient concepts.

As the reader may well imagine, this effort started out as a single E-book covering the field of smart drug-delivery nanovehicles. However, as the work progressed, it became clear that this was a highly active field with new publications coming out in the scientific literature almost every day. Faced with the E-book becoming greatly extended in length, we decided to prepare the subject in two distinct parts. Fortunately, this was not too difficult as there is a natural divide between those stimuli, which can be classified as 'internal' in nature (E-book 1), and those which would be considered 'external' in nature (E-book 2). The internal stimuli comprise those factors which are naturally characteristic of tumors, other disease states, or particular organs or tissues. These stimuli include pH, specific enzymes, redox potential (oxidizing or reducing), and specific biomolecules such as glucose or ATP etc. The external stimuli include those physical energies and forces, which can be applied from outside the body either to guide a nanovehicle to its destination, or to activate it at a specific location once it has arrived. These stimuli include light, temperature (which can be either internal or external), magnetic fields, ultrasound, and electrical and mechanical forces. Dual stimulus and multi-stimuli-responsive systems, and the global market for DDSs are covered in E-book 1, while the important subject of nanotoxicology is covered in E-book 2; subsequently, comprehensive discussions are provided under scrutiny in both E-books.

Acknowledgments

The authors would like to express their gratitude to all who helped them. Special thanks should be given to Professor Michael R Hamblin, for his permanent advice and encouragement of our research into smart nanosystems in nanomedicine and drug/gene delivery systems, and his guidance through the process of writing this book. Second, the authors would like to express their heartfelt gratitude to their beloved families for all their love and encouragement through the years and also while completing this book, to their parents who raised them with a love for science and a conscience, and also to their wives.

The authors deeply extend their appreciation to Seyed Masoud Moosavi Basri and Mahnaz Bozorgomid for composing some of the schematic figures used in this book. Finally, it is our pleasure to acknowledge the guidance and contribution of the Production team at Morgan & Claypool and IOP Publishing, for their expert help.

Author biography

Mahdi Karimi

Mahdi Karimi received his BSc in Medical Laboratory Science from the Iran University of Medical Science (IUMS), in 2005. In 2008, he gained his MSc in Medical Biotechnology from Tabriz University of Medical Science and joined the Tarbiat Modares University as a PhD student in the field of nanobiotechnology. He completed his research in 2013.

During his research, in 2012, he affiliated with the laboratory of Professor Michael Hamblin in the Wellman Center for Photomedicine at Massachusetts General Hospital and Harvard Medical School as a researcher visitor, where he contributed to the design and construction of new smart nanoparticles for drug/gene delivery. On completion of this study, he joined, as Assistant Professor, the Department of Medical Nanotechnology at IUMS. His current research interests include the design of smart nanoparticles in drug/gene delivery and microfluidic systems. He has established a scientific collaboration between his lab and Professor Michael Hamblin's lab to design new classes of smart nanovehicles in drug/gene delivery systems.

Parham Sahandi Zangabad

Parham graduated with a BSc from Sahand University of Technology (SUT), Tabriz, Iran, in 2011. He received his MSc in Nanomaterials/Nanotechnology from Sharif University of Technology (SUT), Tehran, Iran. Concurrently, he became the research assistant at the Research Center for Nanostructured and Advanced Materials (RCNAM), SUT, Tehran, Iran. As a BSc and then MSc student he worked on the assessment of microstructural/mechanical properties of friction stir welded pure copper and friction stir processed hybrid TiO_2–Al_3Ti–MgO/Al nanocomposites. Furthermore, he has done several experiments on synthesis and characterization of sol–gel fabricated ceramic nanocomposite particles.

The advent of innovative nanomaterials and nanotechnology interested him in interfacial sciences/technologies and also nanomedicine, including nanoparticle-based drug delivery systems and nanobiosensors.

He has now joined Professor Karimi's Nanobiotechnology Research lab in the Iran University of Medical Science, Tehran, Iran, in association with Professor Hamblin from Harvard Medical School, Boston, USA; working on smart micro/nanocarriers applied in therapeutic agent delivery systems employed for diagnosis and therapy of various diseases and disorders such as cancers and malignancies, inflammations, infections, etc.

Amir Ghasemi

Amir did his BSc at Sharif University of Technology (SUT), the most prestigious technical university in Iran. He joined the polymeric materials research group in 2012, and received his MSc in Materials Engineering from SUT. For his MSc project, he worked on *thermoplastic starch (TPS)/cellulose nanofibers (CNF) biocomposites*, under the supervision of Professor Bagheri. He synthesized a fully biodegradable nanocomposite, and evaluated the effects of CNF on mechanical and biodegradation of TPS.

His research interests lie in the area of mechanical properties of biopolymers and polymer composites, ranging from material design to the performance of the final product. He also works on micro/nano materials, and bio-based polymers as drug carriers under the supervision of Professor Karimi and Professor Hamblin from the Harvard Medical School.

He now works at Parsa Polymer Sharif, involved in thermoplastics compounding. He would like to thank Professor Karimi and Professor Hamblin for the opportunity to contribute to this work and most importantly learn about such drug delivery systems.

Michael R Hamblin

Michael R Hamblin PhD is a principal investigator at the Wellman Center for Photomedicine, Massachusetts General Hospital, an associate professor of dermatology, Harvard Medical School and the affiliated faculty of Harvard–MIT Division of Health Science and Technology. He directs a laboratory of around 12 scientists who work in photodynamic therapy and low-level light therapy. He has published 274 peer-reviewed articles, is associate editor for eight journals and serves on NIH study sections. He has edited ten proceedings volumes, together with four other major textbooks on PDT and photomedicine. In 2011 Dr Hamblin was honored by election as a Fellow of SPIE.

Smart Internal Stimulus-Responsive Nanocarriers for Drug and Gene Delivery

Mahdi Karimi, Parham Sahandi Zangabad, Amir Ghasemi and Michael R Hamblin

Chapter 1

Introduction

In the current century, the advent of the interdisciplinary/multidisciplinary field of nanobiotechnology (a combination of nanotechnology and biotechnology) has shown great promise for advances in chemistry, physics, bioengineering, nanomedicine and biomedicine, life sciences and therapeutic sciences, and a vast range of new applications are emerging. Nanobiotechnology has provided innovative concepts to improve pharmaceuticals and medicines [1–3], and various biomedical applications are being rapidly developed, including imaging and diagnostics, drug design [4, 5], biosensing and detection [6], tissue regeneration [7], bone regeneration [8], and protein and genome sequencing [9]. In particular, new approaches have emerged for therapies that address formidable diseases, disorders and malignancies. For example, these innovative concepts may provide efficient treatments for various cancers [10] and viral infections, promote effective delivery of peptides, proteins, genes and RNA [11, 12], and aid delivery of therapeutics to the brain [13] by employing nanosystems for the delivery of therapeutic agents.

Synthetic nanoparticles (NP) have fostered tremendous progress in drug delivery systems (DDS), diagnostics and therapeutics [14–16]. Much effort has been invested in the fabrication of new micro/nanocarriers for the transport of various therapeutic agents in biological milieus [17–20], in particular for cancer therapy [21–23]. The concept of smart drug delivery vehicles involves designing and preparing a nanostructure (or microstructure) that can be loaded with a cargo, which can be a therapeutic drug, a contrast agent for imaging, or a nucleic acid for gene therapy. The nanocarrier serves to protect the cargo from degradation by enzymes in the body, enhance the solubility of insoluble drugs, extend the circulation half-life and enhance the penetration and accumulation at the target site. Importantly, smart nanocarriers can be designed to be responsive to a specific stimulus, so that the cargo is only released or activated when desired [24, 25]. Figure 1.1 illustrates several important types of NP used for the design of smart micro/nanocarriers.

doi:10.1088/978-1-6817-4257-1ch1

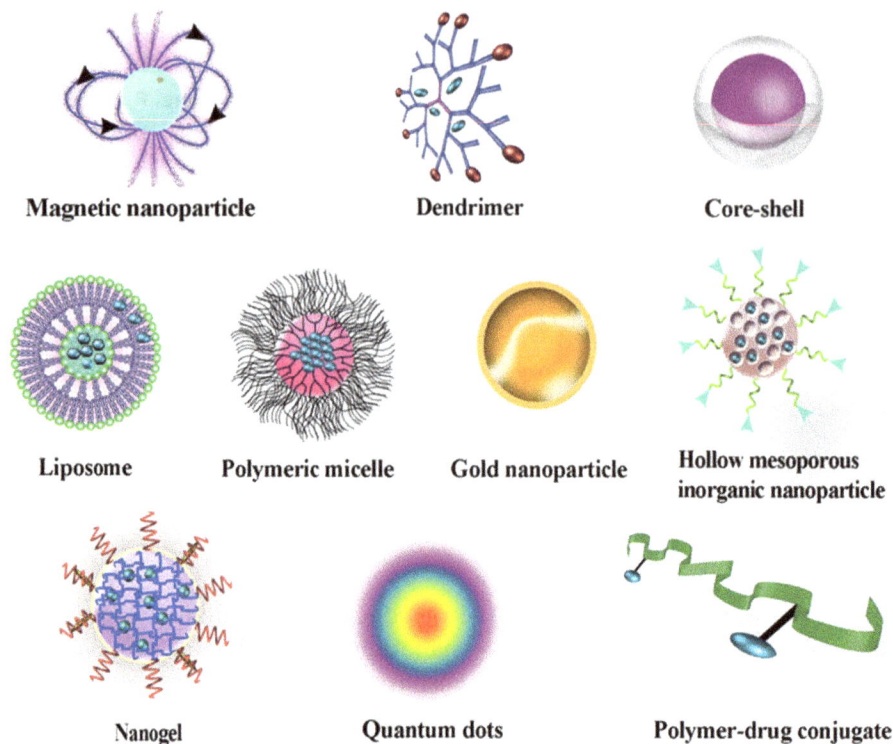

Figure 1.1. Several different types of NP applied in the design of smart micro/nanocarriers.

Drug/gene delivery systems (DGDS) have been designed to respond to a wide range of stimuli, including externally applied stimuli such as magnetic fields, temperature alterations, light irradiation, ultrasound waves, mechanical stress and electrical fields. Internal stimuli, such as pH, biomolecules and redox (oxidative and reductive reactions), can be associated with physico-chemical alterations at specific disease state locations inside the body and have been exploited in recent investigations [25–29].

In internally triggered DGDS, the smart targeting concept comes from the physiochemical phenomena occurring in different biological sites. It is worth noting that the human body is a complex collection of many different environments, each with its own set of physical and chemical parameters. Each cell is surrounded by special receptors on its surface and contains specific enzymes and other molecules to alter its redox potential from that of a normal external body fluid [30, 31]. The various features of different cells and tissues can be taken advantage of to enable the specific targeting and delivery of drugs. Several smart nanocarriers have been reported that can respond to sequences of DNA, have antibody-mediated specificity, or are sensitive to different enzyme reactions. In addition, every individual enzyme possesses its own optimum pH, co-enzymes, temperature and site of action, thus an enzyme-responsive nanocarrier can easily be fabricated with a multi-responsive action. The presence of specific enzymes on the surface of individual cells can specifically determine the uptake of a drug into that cell, while different enzymes present within the cell can control intracellular drug release. In the cancer/tumor environment there are alterations in pH, ATP content, reactive oxygen species (ROS, e.g. H_2O_2), glucose and glutathione concentrations. ROS and reactive nitrogen species (RNS) both exist in many biological microenvironments. For example, high levels of ROS (known to be involved in signaling pathways) are found in cancer cells, suggesting that an oxidizing environment might be a promising target for anticancer drug delivery [32–36]. The highly acidic environment of tumor tissue or the over-expression of a particular enzyme in cancer cells can guide the modification of drug-carrier vehicles so that drug release can be triggered by various stimuli, internal or external [30, 37, 38].

Internally triggered delivery systems can be controlled by taking advantage of the pH value of a specific biological site and its alterations, responding to the redox potentials of different biological sites, or responding to the activity of specific biomolecules, such as enzymes, adenosine-5′-triphosphate (ATP), glucose, glutathione, ROS, etc [25, 35, 39, 40]. Figure 1.2 is a schematic depiction of various internal stimuli that can be applied in smart DGDS.

In the arena of smart micro/nanocarriers triggered by internal stimuli, one innovation is the possible combination of two or more of the aforementioned stimuli with each other or with external stimuli such as light and temperature

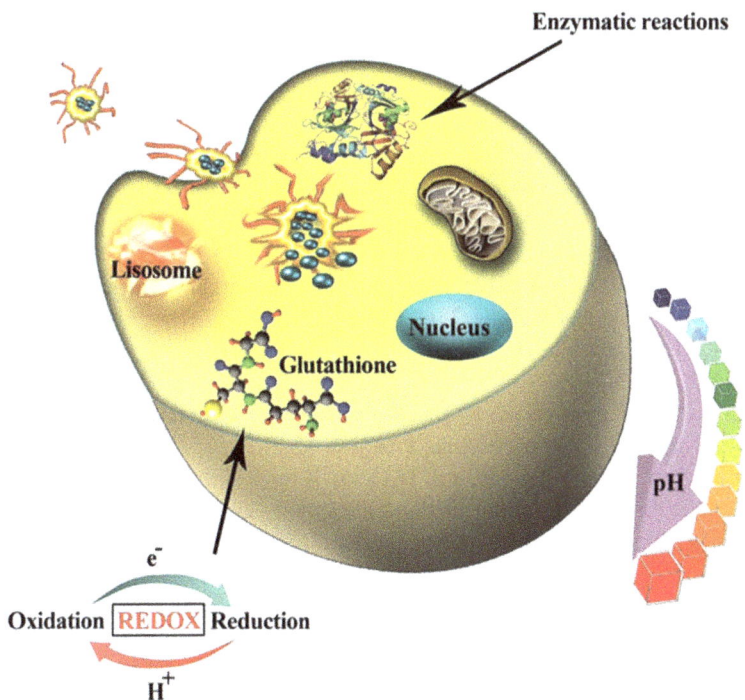

Figure 1.2. Schematic of different classes of internal stimuli, including pH alterations, enzymatic activity and redox potentials, that can act as triggers for the design of smart stimuli-responsive targeted DGDS.

(i.e. dual/triple-stimuli-responsive DGDS), which can result in efficient drug/gene targeting systems [39, 41–44].

This book is concerned with internal stimuli-responsive smart DGDS, involving pH alterations, redox potentials, enzyme activity (the most important biomolecular stimulus in DGDS) and other biomolecular activities/reactions associated with specific biological environments. Furthermore, dual/triple-stimuli-responsive DGDS are discussed in chapter 6. Finally, in chapter 7 future perspectives and a brief assessment of the global market for DDS are presented.

References

[1] Maine E, Thomas V J, Bliemel M, Murira A and Utterback J 2014 The emergence of the nanobiotechnology industry *Nat. Nanotechnol.* **9** 2–5
[2] Morais M Gd, Martins V G, Steffens D, Pranke P and da Costa J A V 2014 Biological applications of nanobiotechnology *J. Nanosci. Nanotechnol.* **14** 1007–17
[3] Logothetidis S 2012 *Nanomedicine and Nanobiotechnology* (Berlin: Springer)
[4] Khare S, Alexander A and Amit N 2014 Biomedical applications of nanobiotechnology for drug design, delivery and diagnostics *Res. J. Pharm. Technol.* **7** 915–25
[5] Doerr A 2014 Nanobiotechnology: imaging without labels *Nat. Method.* **11** 990
[6] Albrecht T 2011 Nanobiotechnology: a new look for nanopore sensing *Nat. Nanotechnol.* **6** 195–6
[7] Chung R-J and Chen C-C 2014 Editorial: nanobiotechnology for tissue regeneration *Curr. Nanosci.* **10** 171–2
[8] Gusić N *et al* 2014 Nanobiotechnology and bone regeneration: a mini-review *Int. Orthop.* **38** 1877–84
[9] Doerr A 2014 Nanobiotechnology: stepping toward sequencing single proteins *Nat. Methods* **11** 611
[10] Li J, Sharkey C, Huang D and King M 2015 Nanobiotechnology for the therapeutic targeting of cancer cells in blood *Cell Mol. Bioeng.* **8** 137–50
[11] Amaral A and Felipe M 2013 Nanobiotechnology: an efficient approach to drug delivery of unstable biomolecules *Curr. Protein Peptide Sci.* **14** 588–94
[12] Karimi M *et al* 2015 Carbon nanotubes part II: a remarkable carrier for drug and gene delivery *Expert Opin. Drug Deliv.* **2015** 1–17
[13] Jain K K 2012 Nanobiotechnology-based strategies for crossing the blood–brain barrier *Nanomedicine* **7** 1225–33
[14] Parveen S, Misra R and Sahoo S K 2012 Nanoparticles: a boon to drug delivery, therapeutics, diagnostics and imaging *Nanomedicine* **8** 147–66
[15] Lee D-E, Koo H, Sun I-C, Ryu J H, Kim K and Kwon I C 2012 Multifunctional nanoparticles for multimodal imaging and theragnosis *Chem. Soc. Rev.* **41** 2656–72
[16] Couvreur P 2013 Nanoparticles in drug delivery: past, present and future *Adv. Drug Deliv. Rev.* **65** 21–3
[17] Collnot E-M, Ali H and Lehr C-M 2012 Nano- and microparticulate drug carriers for targeting of the inflamed intestinal mucosa *J. Controlled Release* **161** 235–46
[18] Yoo J-W, Doshi N and Mitragotri S 2011 Adaptive micro and nanoparticles: temporal control over carrier properties to facilitate drug delivery *Adv. Drug Deliv. Rev.* **63** 1247–56
[19] Wilczewska A Z, Niemirowicz K, Markiewicz K H and Car H 2012 Nanoparticles as drug delivery systems *Pharmacol. Rep.* **64** 1020–37

[20] Karimi M *et al* 2015 Carbon nanotubes part I: preparation of a novel and versatile drug-delivery vehicle *Expert Opin. Drug Deliv.* **2015** 1–17

[21] Brannon-Peppas L and Blanchette J O 2012 Nanoparticle and targeted systems for cancer therapy *Adv. Drug Deliv. Rev.* **64** 206–12

[22] Hu C-M J and Zhang L 2012 Nanoparticle-based combination therapy toward overcoming drug resistance in cancer *Biochem. Pharmacol.* **83** 1104–11

[23] Zhang X-X, Eden H S and Chen X 2012 Peptides in cancer nanomedicine: drug carriers, targeting ligands and protease substrates *J. Controlled Release* **159** 2–13

[24] Fleige E, Quadir M A and Haag R 2012 Stimuli-responsive polymeric nanocarriers for the controlled transport of active compounds: concepts and applications *Adv. Drug Deliv. Rev.* **64** 866–84

[25] Mura S, Nicolas J and Couvreur P 2013 Stimuli-responsive nanocarriers for drug delivery *Nat. Mater.* **12** 991–1003

[26] Duan Q *et al* 2013 pH-responsive supramolecular vesicles based on water-soluble pillar [6] arene and ferrocene derivative for drug delivery *J. Am. Chem. Soc.* **135** 10542–9

[27] Radhakrishnan K, Tripathy J and Raichur A M 2013 Dual enzyme responsive micro-capsules simulating an 'OR' logic gate for biologically triggered drug delivery applications *Chem. Commun.* **49** 5390–2

[28] Huo M, Yuan J, Tao L and Wei Y 2014 Redox-responsive polymers for drug delivery: from molecular design to applications *Polymer Chem.* **5** 1519–28

[29] Cheng R, Feng F, Meng F, Deng C, Feijen J and Zhong Z 2011 Glutathione-responsive nano-vehicles as a promising platform for targeted intracellular drug and gene delivery *J. Controlled Release* **152** 2–12

[30] Wu L, Zou Y, Deng C, Cheng R, Meng F and Zhong Z 2013 Intracellular release of doxorubicin from core-crosslinked polypeptide micelles triggered by both pH and reduction conditions *Biomaterials* **34** 5262–72

[31] Cheng R *et al* 2011 Reduction and temperature dual-responsive crosslinked polymersomes for targeted intracellular protein delivery *J. Mater. Chem.* **21** 19013–20

[32] Wu H *et al* 2014 Prostate stem cell antigen antibody-conjugated multiwalled carbon nanotubes for targeted ultrasound imaging and drug delivery *Biomaterials* **35** 5369–80

[33] Yu J *et al* 2015 Microneedle-array patches loaded with hypoxia-sensitive vesicles provide fast glucose-responsive insulin delivery *Proc. Natl Acad. Sci.* **112** 8260–5

[34] Joshi-Barr S, de Gracia Lux C, Mahmoud E and Almutairi A 2014 Exploiting oxidative microenvironments in the body as triggers for drug delivery systems *Antioxid. Redox Signal.* **21** 730–54

[35] Zhu C-L, Wang X-W, Lin Z-Z, Xie Z-H and Wang X-R 2014 Cell microenvironment stimuli-responsive controlled-release delivery systems based on mesoporous silica nano-particles *J. Food Drug Anal.* **22** 18–28

[36] Vatansever F *et al* 2013 Antimicrobial strategies centered around reactive oxygen species–bactericidal antibiotics, photodynamic therapy, and beyond *FEMS Microbiol. Rev.* **37** 955–85

[37] Alvarez-Lorenzo C, Bromberg L and Concheiro A 2009 Light-sensitive intelligent drug delivery systems *Photochem. Photobiol.* **85** 848–60

[38] Chan A, Orme R P, Fricker R A and Roach P 2013 Remote and local control of stimuli responsive materials for therapeutic applications *Adv. Drug Deliv. Rev.* **65** 497–514

[39] Cheng R, Meng F, Deng C, Klok H-A and Zhong Z 2013 Dual and multi-stimuli responsive polymeric nanoparticles for programmed site-specific drug delivery *Biomaterials* **34** 3647–57

[40] You J-O, Almeda D, George J and Auguste D T 2010 Bioresponsive matrices in drug delivery *J. Biol. Eng.* **4** 1–12

[41] Huang X *et al* 2013 Triple-stimuli (pH/thermo/reduction) sensitive copolymers for intracellular drug delivery *J. Mater. Chem.* B **1** 1860–8

[42] Han D, Tong X and Zhao Y 2012 Block copolymer micelles with a dual-stimuli-responsive core for fast or slow degradation *Langmuir* **28** 2327–31

[43] Dong J *et al* 2013 Multiple stimuli-responsive polymeric micelles for controlled release *Soft Matter* **9** 370–3

[44] Cheng Z, Al Zaki A, Hui J Z, Muzykantov V R and Tsourkas A 2012 Multifunctional nanoparticles: cost versus benefit of adding targeting and imaging capabilities *Science* **338** 903–10

Chapter 2

pH-sensitive micro/nanocarriers

2.1 Introduction

pH-responsive nanostructures have been applied in various technologies, including pH-sensors [1], theranostic applications [2, 3], DGDS, etc. In addition, pH-responsive smart nano/micro particles can be utilized in controllable switches, controlled-release surfaces, controllable wettability and the specific recognition of cells [4].

There are distinct pH differences in various tissues and cellular compartments of the human body, and pH gradients have also been shown in both healthy states and pathological conditions. For example, there are pH gradients in the sub-cellular environment, with pH differences between the lysosomes (4.5–5), endosomes (5.5–6), cytosol (7.4), Golgi apparatus (6.4), etc [5]. The pH changes seen along the gastrointestinal tract (GIT) are an example of such pH alterations in biological environments. Furthermore, the growth of microorganisms can change the pH of their immediate environment, either directly or through inducing the release of enzymes from the host defense. Healing and non-healing wounds can also create acidic and alkaline milieus, respectively [6].

Smart DGDS can release their drug or gene load in response to the stimulus of a pH change between normal physiological pH (i.e. 7.4) and the lower pH found in cancer. The pH profiles in cancerous tissues differ from those in normal tissues. Because of the rapid proliferation of cancerous cells, the vascular system in cancerous tissues is often inefficient; consequently, the tissues do not receive adequate supplies of oxygen and nutrients. Under anaerobic conditions, cancerous tissues carry out glycolysis (the Warburg effect) rather than oxidative phosphorylation, and therefore produce lactic acid, which makes their environment more acidic, leading to an extracellular pH value that is lower than the normal blood pH (i.e. 7.4) [7, 8]. In fact, lactic and carbonic acid are the main two sources of low pH in

doi:10.1088/978-1-6817-4257-1ch2
2-1

the tumor mass [9]. According to the Warburg effect, the preference for glucose fermentation (glycolysis) remains even in the presence of adequate oxygen and some experts believe it to be a cause rather than a result of cancer [10, 11]. Furthermore, the pH of cancer cells varies depending upon the cancer type, size and location [12]. Sometimes the environmental pH increases because necrosis occurs in the large tumor mass [12]. Moreover, in addition to the lower extracellular pH of tumor tissues, in the intracellular environment the adenosine triphosphate (ATP)-driven H^+ pump (located in the endosomal membrane) pulls H^+ ions into the endosomal lumen from the cytosol and further reduces the pH [13].

This lower pH can cause a number of problems in cancer treatment. For example, the low pH causes reduced activity in many common anticancer drugs, such as doxorubicin (DOX) [14]. Overcoming such challenges requires a major effort to find new solutions.

pH-responsive therapeutic delivery systems enable researchers to find solutions for the aforementioned issues, while utilizing the highly efficient targeting capability of such pH-triggered systems for the selective delivery of therapeutic cargos.

2.2 pH-sensitive nanocarriers

The pH-sensitive nanomaterials used in DGDS can be classified as organic, inorganic or hybrid according to their constituents [15]. Various nanomaterials have been evaluated as pH-sensitive nanocarriers. A summary of the various types of pH-responsive nanocarriers (including polymeric micelles, liposomes, nanogels, core–shell NP, polymer–drug conjugates and inorganic NP) and their effects on DGDS is given in table 2.1.

2.3 pH-sensitive micro/nanocarrier drug release mechanisms

Utilizing pH-sensitive micro/nanocarriers for targeted drug delivery applications requires that their characteristics and mechanisms be fully understood. Hence, many new pH-responsive/gene delivery systems that use innovative nanomaterials have been developed in recent years. Drug release in pH-responsive NP is effected by various mechanisms, including dissolution of the nanocarrier [47], swelling of the polymeric nanocarrier [48, 49], or both dissolution and swelling [50] at a specific pH using acid-swellable groups [51].

Consequently, understanding the pH-responsiveness mechanisms of these systems is of great importance. Several varieties of NP have been confirmed to have pH-sensitivity, including polymeric particles, liposomes, micelles, dendrimers, etc. In pH-responsive polymeric particle-based DDS, the drug is conjugated to polymeric particles through pH-responsive spacers. These spacers can be degraded within the low environmental pH of tumors or endosomes/lysosomes [52]. Therefore, the pH-responsiveness of polymeric carriers can be varied through several factors, such as the hydrophobic carbon chain length or monomer type and the proportion of different constituent monomers [53, 54]. In liposomes, internalization of the modified liposomes into cells occurs through endocytosis into endosomes and the subsequent fusion with lysosomes. The cargo is then

Table 2.1. Examples of different kinds of pH-responsive nanocarriers.

Nanocarrier type	Particle synthesis method	Particle characteristics	pH-dependent therapeutic effect or outcome	Reference
Polymeric micelles	Self-assembled spherical supramolecular aggregates formed by amphiphilic block copolymers	• Capable of solubilizing insoluble drug molecules; enhanced solubility; reduced toxicity; the enhanced permeability and retention (EPR) effect for passive targeting • Weakly water-soluble drug molecules can be encapsulated in the core of micellar structures	• Prevention of drug release at physiological pH but facilitated drug release at endosomal pH (pH 5.0) • Tumor infiltration permeability, effective antitumor activity and low toxicity • Increased serum stability of the nanocarriers in the blood • Avoidance of premature drug release	[16–20]
Liposomes	Self-assembled spherical vesicles made up of lipid bilayer structures	• Can be used as carriers of both hydrophobic and hydrophilic molecules • Improved control over size, surface charge and targeting ability toward diseased sites • Better compatibility than micelles • Limitations such as low encapsulation effectiveness; fast drug release rate; low storage stability; lack of tunable triggers for drug release	• Improved targeting and drug release by surface modification and utilization of layer-by-layer liposome NP • Enhanced gene targeting in the liver followed by reduction in plasma cholesterol • Multi-drug pH-sensitive delivery capability with controlled and different drug release rates and improved biodistribution profile • Antigen delivery • Vaccine delivery • Increased cellular uptake of DOX-loaded liposomal carrier at pH 6.4	[18, 21–28]
Polymeric nanogels	Highly porous three-dimensional network created through self-assembly or covalent bonding of crosslinked hydrophilic polymer chains	• Higher drug-loading capacity than micelles and liposomes (except for hydrophobic drugs); efficient targeted delivery	• Biocompatible pH-responsive nanogels capable of loading hydrophobic drugs • Dual pH/temperature-responsive hyperthermia therapy for cancer cells with increased therapeutic drug activity	[29–32]

(*Continued.*)

Table 2.1. (Continued.)

Nanocarrier type	Particle synthesis method	Particle characteristics	pH-dependent therapeutic effect or outcome	Reference
Polymer–drug conjugates	Polymeric chains	• Drugs covalently conjugated to pH-sensitive polymeric chains • High blood circulation time	• Increased water solubility for drugs; tumor targeting • High sensitivity, with a strong radio-sensitizing effect • Capable of loading poorly soluble drugs	[33–35]
Core–shell NP	Polymeric colloidal particles with spherical, branched and core–shell forms	• Core–shell NP can be prepared via synthetic or natural polymers	• Core–shell drug carriers with anti-tumor ability; faster drug release at lower pH • Capable of delivering protein • Notable hypoglycemic effects and enhanced insulin-relative bioavailability • Highly sensitive dual pH/temperature responsiveness; sharp tunable phase transitions of dual temperature and pH-responses	[36–40]
Inorganic NP	Ceramic (e.g. MSN) or metal NP (e.g. gold) with functional pH-sensitive polymeric moieties	• Controllable surface functionalizaton with good encapsulation efficiency for drugs/genes	• Increased cytotoxicity through enhanced cellular uptake mechanisms • In MSN, blocked pores at neutral pH and drug release in acidic conditions; low cytotoxicity and improved biocompatibility • Good drug release and targeting capability • Good drug loading capability	[36, 41–46]

delivered to the cytoplasm [26]. The pH-sensitivity of micelles is based on the protonation of moieties such as carboxylic groups or titratable amines attached to the surface of the polymeric particle. The form of these particles changes as the pH change occurs in the environment [55]. In dendrimers, one important pH-responsive method for triggered drug release is via the particle cleavage that occurs when hydrophobic groups, sensitive to acids, are used [56]. pH-responsive lipid-based delivery systems utilize nanocarriers such as liposomes, acid-labile zwitterionic peptide lipid derivatives, etc [57]. The use of the instability of different pH-responsive nanocarriers in biological media is based on conventional or innovative nanomaterials utilizing ionizable pH-sensitive groups (e.g. in polymeric and peptide NP) and acid-labile chemical bonds, which will be discussed.

2.3.1 Ionizable pH-sensitive nanocarriers

Polymeric NP

Ionizable chemical moieties such as amines, phosphoric acids and carboxylic acid are able to accept or donate protons in response to pH alteration. Hence, these chemical moieties can be used in polymeric NP. The features, preparation and application of pH-sensitive polymers have been reviewed [58]. Polymeric NP are categorized into two major groups based on their charges: anionic polymers and cationic polymers. It is known that pH-triggered polymers contain weakly acidic or weakly basic chemical moieties. During a pH change, the weak acid serves as a proton donor and its conjugated base is a proton acceptor. Carboxylic acids are weak acids that are used in anionic polymers, including poly(acrylic acid) (PAA), poly(butylacrylic acid) (PBAA), poly(propylacrylic acid) (PPAA), poly(ethylacrylic-acid) (PEAA), poly(methylacrylic acid) (PMAA), etc [59–62].

In one study, PAA polymers were utilized to fabricate hybrid silver–polyacrylic acid–poly(N-vinylpyrrolidone) (Ag–PAA–PVP) nanogels. The results indicated that the pH-sensitive hybrid nanogels can be considered for biomedical applications [63]. The sulfonamide moiety is another group that is applicable in anionic polymers; initially, it was used in an antibacterial compound for infection therapy. Figure 2.1 shows its chemical structure. As the hydrogen atom of the amide nitrogen (N^1) can easily ionize and donate a proton, sulfonamide can serve as a weak acid [64, 65].

Cationic polymers increase cellular uptake because of their positively charged surface [66]. Amino-containing polymers release the loaded gene or drug cargo molecules via the proton sponge effect [67–69]. Cationic polymers acting in line with the proton sponge effect have displayed enhanced delivery efficiency for therapeutics such as nucleic acids, drugs [70] and proteins [71, 72]. The cationic polymers poly(L-lysine) (PLL) (widely used for DNA delivery) [73] and polyethylenimine (PEI) (in both linear and branched forms) offer highly efficient delivery of oligonucleotides [74], plasmid DNA (pDNA) [75, 76] and RNA such as siRNA [77, 78].

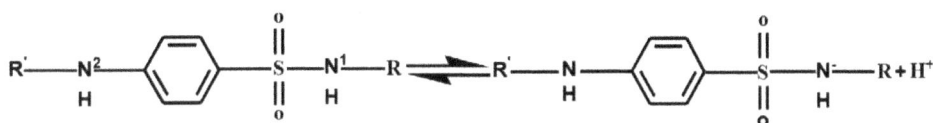

Figure 2.1. Chemical structure of sulfonamide.

Furthermore, pH-responsive nanocarriers have been shown to cause facilitated endosomal membrane rupture through the proton sponge effect, in addition to other endosome destabilization mechanisms such as surface charge reversal of the nanocarrier and deshielding (i.e. active ligand exposure at the outer surface of the nanocarrier) at the extracellular pH of the tumor milieu [79].

Chitosan is considered to be a natural cationic aminopolysaccharide derived from the partial deacetylation of chitin and it is a copolymer of N-acetylglucosamine and glucosamine linked by a 1–4 glycosidic linker. Sung *et al* reviewed advances in macromolecular delivery via chitosan-based NP [80]. Chitosan is nontoxic and biodegradable and is widely applied in DDS for the controlled release of proteins (e.g. insulin) [81] and for controlled oral delivery, and it is also used for the entrapment of enzymes [82], and it is optimized for gen delivery [83, 84]. The significant properties of chitosan and its derivatives, including adhesion to mucosal surfaces and the ability to penetrate between tight junctions in epithelial cells, make them favorable candidates for the oral delivery of therapeutics [85, 86]. Various chitosan-based nano-assembled NP are illustrated in figure 2.2. In one study, a pH-triggered polyelectrolyte 'complex sandwich' microparticle was fabricated from alginate/oligochitosan/Eudragit(®) L100-55. The results showed the pH-sensitive release of drugs in conditions that simulated the intestinal environment (pH 6.8) [87]. In another study, Jahromi et al [88] showed that the maturation of dendritic cells

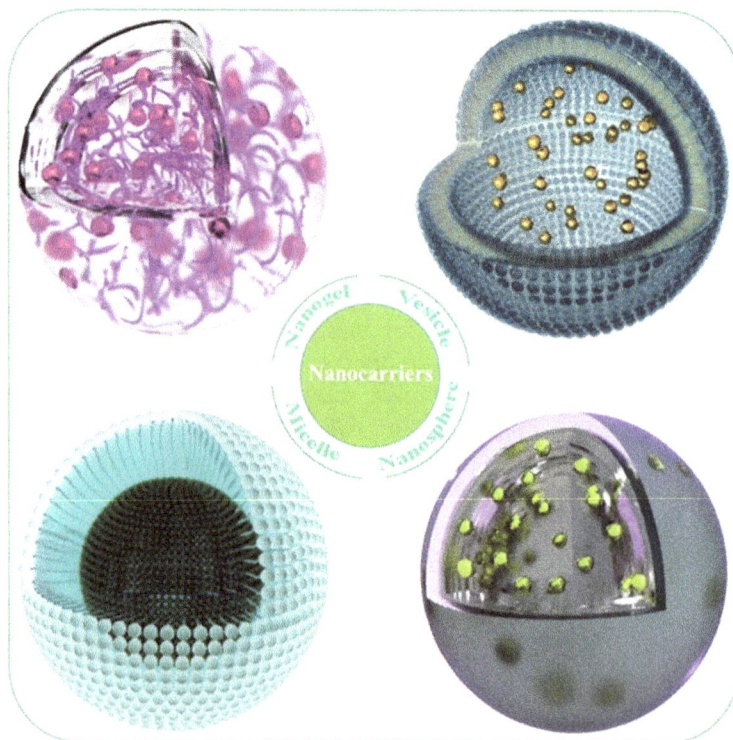

Figure 2.2. Chitosan-based nanoassemblies applied for biomedicine including nanosphere, nanovesicle, micellar structure and nanogel.

(DCs) is induced by chitosan-tripolyphosphate (TPP). Their results demonstrated that chitosan-TPP treats DCs in an efficient manner.

The self-assembly behavior of chitosan is enabled by the presence of functional moieties, including –OH and –NH_2. Progress in the synthesis of self-assembled chitosan-based NP such as nanoshells, nanocomplexes and micelles as systems for the delivery of drugs, genes and small molecules has been reviewed [89]. Nogueira *et al* developed methotrexate-loaded chitosan NP produced using a modified ionotropic complexation process, and these led to apoptotic effects in tumor cells [90]. In addition, nucleic acids can be delivered via chitosan-based NP such as polyaspartamide (PASPAM) [91], poly(malic acid) [92, 93] and poly(histidine) [19]. In a study by Cheng *et al*, a pH-triggered and non-viral gene delivery system was prepared. These NP showed low toxicity to HepG2 and KB cells, and although it was possible to enhance gene transfection and expression in KB cells, free folic acid (FA) inhibited this process significantly [94].

Mesoporous silica NP (MSN) have shown great potential in stimuli-responsive DGDS, especially pH-sensitive nanocarriers. Wen *et al* [95] reported the synthesis of pH-responsive composite microspheres via the distillation precipitation polymerization process. This nanocarrier was composed of a Fe_3O_4 nanoparticle core, a triple layer of MSN and a cover of crosslinked PMAA as a shell in which the DOX model drug was loaded. The results indicated that the cumulative release rate of the DOX from the microsphere was significantly higher below its pKa than above its pKa. Furthermore, pH-sensitive supramolecular nanovalves functionalized on the surface of MSN pores have been utilized in smart delivery systems. Nanovalve-based nanocarriers have been widely studied [95–98]. Nguyen *et al* [97] capped the pores of coumarin 460-loaded mesoporous silica MSM-41 with dibenzo[24]crown-8 (DB24C8) nanovalves; coumarin 460 molecules were then released controllably from the pores, triggered by various bases.

Tao *et al* showed that a pH-sensitive gel matrix can be formed by adding metformin hydrochloride (an antidiabetic drug) to graphene oxide (GO) at room temperature. The N-containing functionalities of the drug formed hydrogen bonds with the hydroxyl and carboxyl groups on the 2D surface of GO sheets [99]. Covalent functionalization of GO by poly(2-(diethylamino) ethyl methacrylate) (PDEA) was used to produce a pH-sensitive delivery vehicle for camptothecin [100]. Ren *et al* formed a stable dispersion of hydrophobic graphene sheets in water using polyacrylamide. This suspension showed a reversible pH response that could switch dispersed graphene sheets to aggregated sheets of grapheme [101].

Several pH-sensitive anticancer DDS, including doxorubicin-loaded dual polymer-coated nanoscale graphene, show remarkable features, such as the reduction of cancer cell resistance to therapeutic drugs, rapid anticancer drug release and slow efflux, while synergistic therapeutic effects against tumor sites can be achieved [102].

Peptides

Peptides are organic molecules that can be used for highly efficient delivery of genes and drugs. pH-sensitive fusogenic peptides, when exposed to the low pH in the endosome, undergo fusion with the endosomal membrane. There are two broad categories of gene delivery vectors: viral and non-viral [103]. Viruses attach to host cells and deliver their genetic materials into the replication cycle of the host cells [104]. Although there have been many studies to reveal the underlying mechanisms through which viruses deliver their genes, intensive research is still needed to assess their true potential as clinical gene delivery systems. *In vitro* and *in vivo* studies of viral vectors have had many successes; however, their viral nature limits their clinical applications. Viral vectors comprise: (1) retroviruses [105]; (2) adenoviruses [106, 107]; (3) adeno-associated viruses [108, 109]; and (4) *Herpes simplex* viruses [110]. Several important side-effects, namely inflammation, sustained immune reconstitution, development of cancer and even death [111–113] have been reported to occur after gene therapy via viral vectors. Although non-viral vectors based on peptides are considered to be less efficient than viruses, their development is motivated by a desire for safer gene delivery.

Subbarao *et al* [114] (1987) created the first peptide Glu–Ala–Leu–Ala (GALA) motif, composed of 30 repeated units of these four amino acids, and this had the specific ability to fuse with plasma membranes. The GALA motif can be added to the liposome surface to increase its efficiency in DGDS. At neutral pH, glutamic acid residues are deprotonated and they become negatively charged as a result. Consequently, these negative charges create repulsions between the carboxylic acid moieties of glutamic acids in GALA and destabilize its alpha helix structure. As the pH decreases, these residues become protonated and cause stabilization of the alpha helix structure of GALA and increase its hydrophobicity, which in turn increases its interaction with cellular lipid bilayers [115].

The large size of GALA reduces its blood circulation time. To overcome this issue, a short form of GALA, called shGALA, was developed by Sakurai *et al* and modified with polyethylene glycol (PEG) [116]. The long circulation time of shGALA and its efficient tumor inhibition and gene delivery distinguished it as a promising carrier in drug delivery applications. However, because of the overall negative charge of the glutamic acids in GALA, it cannot directly bind to nucleic acids. Wyman *et al* designed another kind of peptide family, a cationic amphipathic fusogenic one called KALA(WEAKLAKALAKALAKHLAKALAKALKACEA). This family can bind to DNA and mediate DNA transfection [117]. Interestingly, Guo *et al* developed the KALA peptide as an efficient gene vector. Their investigation showed that an increase in the number of histidine and lysine residues increases the ability of the peptide to bind strongly to DNA. Their results also showed that the oligomerized peptide, tri-RC29 (RGDN3K6H3CKHLAKALAKALAC), had better gene expression and, moreover, the RC29/DNA complexes were less cytotoxic than the PEI/DNA complexes [118].

McCarthy *et al* recently developed a bio-inspired peptide for gene delivery, which was composed of RALA units, with arginine replacing lysine. This study mentioned that arginine is found at a higher rate than lysine in naturally occurring DNA binding/condensing motifs and that poly-arginine has been shown to be a transfection agent. Because sequences that are arginine-rich showed improved cell internalization, this group tested a peptide termed RALA. Their results showed that RALA can be an effective DNA delivery platform and that it has potential for clinical application [119].

2.3.2 Acid-labile chemical linkers

Acid/alkali-labile chemical linkers are an important group of pH-responsive DDS. The development of new acid/alkali-labile linkers enables us to design pH-responsive DDS with high sensitivity.

Diverse acid-labile covalent linkages can be hydrolyzed immediately in acidic environments. Several cleavable linkers include the aforementioned functionalities, e.g. acetal/ketal, orthoester, hydrazone, imine, cis-acotinyl bonds, oximes, etc. The structures of a number of these acid-sensitive bonds are given in table 2.2. Polymer particles comprising these bonds have stability at physiological pH, but drug cargos can be released through the hydrolysis that occurs in response to decreased pH values within biological environments such as intracellular compartments (including endosomes, i.e. early endosomes with a pH of 6.0–6.5 or late endosomes with a pH of 5.0–6.0, and lysosomes with a pH of 4.5–5.0) or via low extracellular pH of solid tumor tissues. Figure 2.3 illustrates an acid-labile nanocarrier conjugated with a drug molecule that enters a cell by endocytosis and releases the drug molecule inside an acidic cellular compartment (either endosomes or lysosomes) through hydrolytic cleavage and detachment of drug from the nanocarrier. It is notable that the cleavable linkers can be placed in the side-chains or in the backbone of the polymer. For example, with regard to amphiphilic copolymers it is suggested that the drug molecules can be conjugated to the side-chains via acid-labile linkers (e.g. the conjugation of hydrophobic anticancer drugs to the hydrophobic segments of an amphiphilic micellar structure) [120, 121].

Heller *et al* [122] proposed the application of polyacetals as drug-releasing systems in 1979. A facile route for the synthesis of polyacetals is via the reaction between divinyl ethers and polyols, as shown in figure 2.4.

Table 2.2. Chemical formulas for several acid-sensitive linkers.

Imine	
Hydrazone	
Oxime	
Ketal	
Acetal	

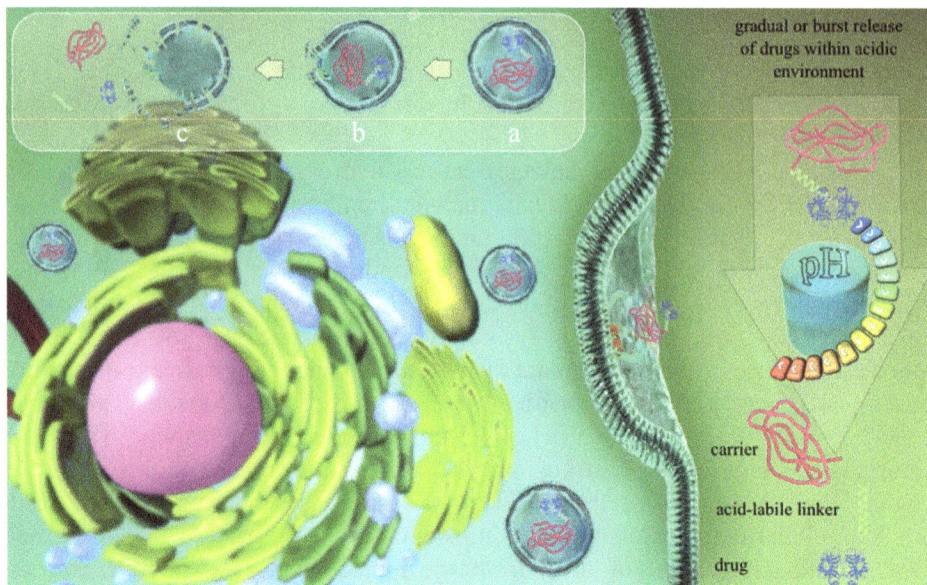

Figure 2.3. Acid-sensitive linkers are used to attach the drug to the polymer blocks and after a pH decrease the structure destabilizes, leading to drug release. Release of drugs induced via an intracellular lower pH value is also indicated.

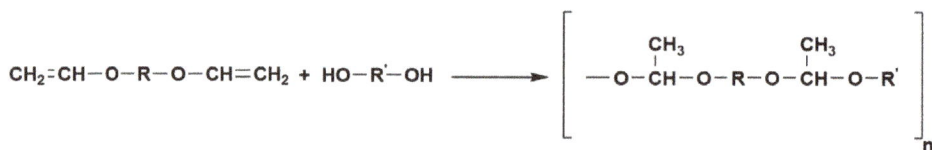

Figure 2.4. Schematic illustration of the reaction between divinyl ethers and polyols.

In recent years, many studies have been conducted on pH-triggered polymeric micro/nanocarriers that contain dacetal/ketal linkers [123]. Non-toxic products (including aldehydes, alcohols and ketones) can be generated by the exposure of acetal- or ketal-containing polymers to acidic environments. Shenoi *et al* designed a ketal-containing multi-functional nanoparticle. This biodegradable and water-soluble polymer comprised hyper-branched polyglycerols in the main chain connected to various ketal structures. It has been suggested that such multifunctional polymers are good candidates for DGDS because of advantages such as their *in vivo* degradability and biocompatibility [124].

Gillies *et al* [125] reported on acid-labile copolymer micelles based on the coupling of hydrophobic moieties to one block of a diblock copolymer via a cyclic benzylidene acetal as an acid-labile linker. A small reduction in pH destabilized the acid-labile linkers and induced the hydrophobic block to turn hydrophilic (as a result of hydrolysis), which then destabilized the entire micellar structure and led to drug release.

Acid-labile ketal-containing polymeric non-viral systems have been employed for the intracellular delivery of nucleic acids to overcome limitations such as the endosomal escape of nucleic acids. Lee *et al* recently prepared ketal-containing poly (β-amino ester) (KPAE) for delivery of siRNA. The ketal linkage can be cleaved in the acidic environment of endosomal compartments. In addition, the buffering capacity of KPAE can condense siRNA into nanocomplexes (150 nm in size) that are immediately dissociated at acidic pH (compared to their high stability at neutral pH), leading to siRNA release. The results indicated that KPAE delivered the loaded siRNA cargo effectively and facilitated endosome disruption through the protein 'sponge effect' and the colloidal osmotic mechanism. Furthermore, the tumor necrosis factor-α (TNF-α) siRNA was delivered to lipopolysaccharide (LPS)-stimulated macrophages, leading to remarkable inhibition of TNF-α expression [126]. Shim and Kwon produced another KPAE. The cationic, hydrolytic and degradable KPAE was synthesized via the polymerization of conjugated diaminoethane followed by the addition of acid-labile amino ketal branches. Within the cytoplasm, the ketal branches of KPAE were hydrolyzed, inducing degradation of the polyester backbone. In addition, DNA/KPAE polyplexes were reported to be efficient serum-resistant gene transporters of DNA into the cytoplasm, with insignificant cytotoxicity [127].

Figure 2.5 shows three-layered self-assembled micelles possessing pH-sensitive properties for the release of hydrophobic anticancer drugs. The micelles swell at acidic pH values due to the ionization of tertiary amine groups in the middle layer. The rate of the drug release was improved by the decrease of the pH value from 7.4 to 5.0. Furthermore, the *in vitro* cytotoxicity of anticancer drug-encapsulated micelles to HepG2 (hepatocellular carcinoma) cells illustrated that the

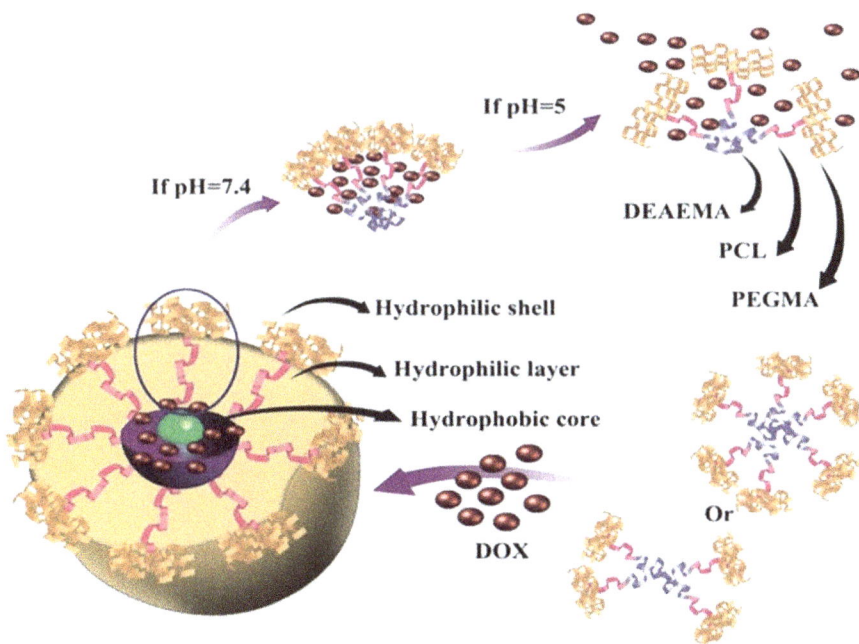

Figure 2.5. Schematic illustration of the loading and release of drug molecules in a pH-triggered micellar structure.

4/6AS-PCL-b-PDEAEMA-b-PPEGMA enhanced anticancer activity as well as the bioavailability of the anticancer drug. Therefore, a lower quantity of the drug can be utilized for anticancer therapeutic applications [128].

Hydrazone moieties are another important group of acid-labile chemical linkers, and are the product of the reaction of hydrazine with aldehydes or ketones [126]. Hydrazone bonds can be utilized in pH-responsive NP for drug delivery applications, particularly in cancer therapy [130–135]. Ding *et al* designed a smart all-in-one nanosystem that carries various favorable functions in a single carrier. Therein, a multifunctional multiblock polyurethane (MMPU) was synthesized composed of pH-responsive hydrazone bonds (HPCL) containing soft segments (to accommodate the lipophilic agents in physiological conditions and trigger release of the drugs in response to an intracellular acidic environment) and biocompatible polyurethane (PU)-based hard segments (figure 2.6(a)) [136].

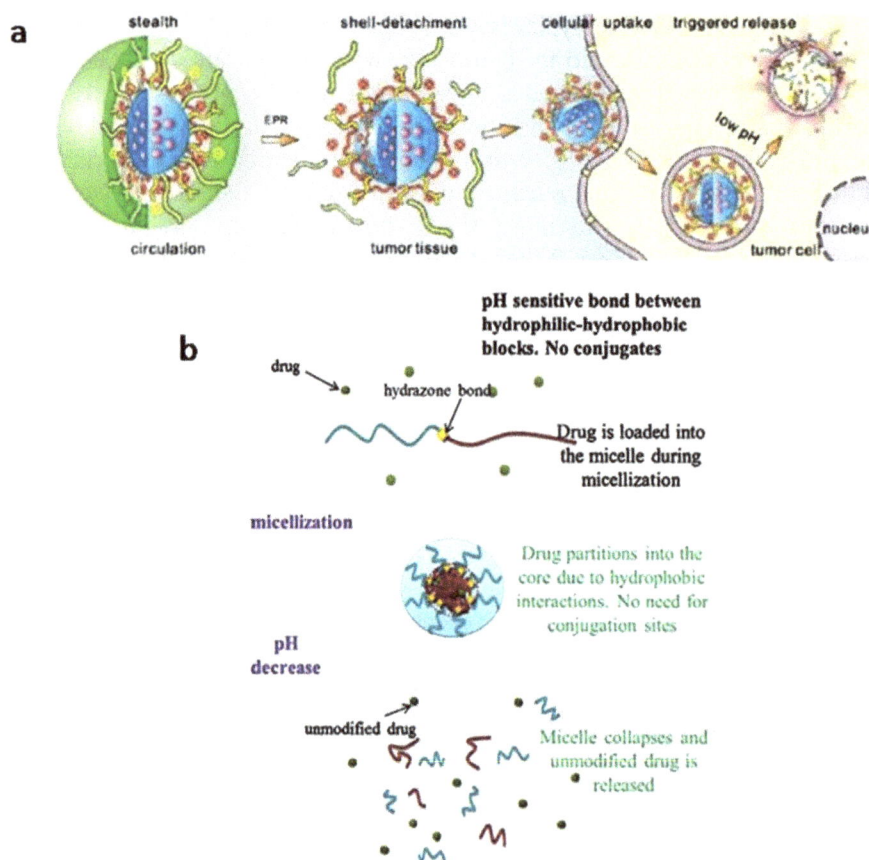

Figure 2.6. (a) Drug delivery and release from a multi-functional MMPU-based nanocarrier towards tumor cells. Reproduced with permission from [133]. Copyright 2013 by The American Chemical Society. (b) The preparation process for diblock copolymer micelles composed of PEG and PCL linked with a pH-labile hydrazone bond followed by drug release via pH reduction. Reproduced with permission from [137]. Copyright 2013 Elsevier.

Controlled and stepwise drug-release was achieved by Zhou *et al* (2012) [137] using a multifunctional and biodegradable polymer as an anticancer drug carrier. They prepared a pH-sensitive PU soft segment together with a hard segment and used hydrazone-linked methoxy-poly(ethylene glycol) (m-PEG-Hyd) as an end-capper. The resulting pH-sensitive polyurethane was shown to be a promising non-toxic and biodegradable multifunctional carrier for controllable and active intracellular drug delivery.

Acid-labile hydrazone bonds comprising micelles have also been reported in recent years for cancer therapy [138]. Wang *et al* [139] fabricated a micellar DDS based on a poly(amino acid)-based amphiphilic copolymer. They used poly(ethylene glycol)-monomethylether (mPEG) and then conjugated DOX to polyaspartic hydrazide (PAHy), prepared by the hydrazinolysis of the poly(succinimide) (PSI), to provide an amphiphilic polymer (PEG-hyd–PAHy-DOX)) with acid-labile Hyd bonds. They showed that the release of DOX at an acidic pH (5.0) was faster than at a physiological pH (7.4). Koutroumanis *et al* [140] developed a diblock copolymer comprising a PEG and a polycaprolactone (PCL) segment linked with a pH-labile hydrazone bond. Micelles formed by this copolymer offered the advantage of encapsulating the highly hydrophobic α-tocopherol drug without the need for conjugation sites (figure 2.6(b)). A multifunctional nanocarrier based on PEI–HZ–DOX (PhD) and PEI–PEG–folate (PPF) was developed for the delivery of drugs and siRNA. DOX was attached to PEI via a hydrazone bound. The results showed that DOX and siRNA were released synchronously in cancer cells via a pH-triggered process, resulting in improved accumulation of siRNA and DOX in the tumor and reduced non-specific targeting because of the combination of pH-stimulated release, the EPR effect of the nanocarriers, the synergistic effect of DOX and siRNA, and folate-mediated targeting [134].

Recently, orthoester chemical linkers have been utilized in amphiphilic copolymer micelles in order to design pH-responsive nanocarriers in DDA for anticancer applications [141]. Tang *et al* designed a copolymer micelle for the pH-responsive delivery of water-insoluble anticancer drugs. These micelles were constructed through the self-assembly of a diblock amphiphilic copolymer with a hydrophilic PEG block and a hydrophobic polymethacrylate block bearing acid-sensitive orthoester side-chains [142].

Furthermore, imine linkers acting as acid-labile moieties were investigated in anticancer delivery systems. Ding *et al* [143] reported that multifunctional drug carriers (e.g. polymeric micelles) can be formed by the pH-responsive progressive hydrolysis of a novel, weak acid-labile benzoic-imine linker in micelle-forming amphiphilic polymers. While the micelles were stable at physiological pH, they partially hydrolyzed at the extracellular pH of the solid tumor and completely hydrolyzed at the even lower endosomal pH. Furthermore, due to the generation of amino moieties from the cleavage of the imine bond at tumor pH, the surface feature of the micelles converted from neutral to positively charged. The cellular uptake of the micelles was facilitated by the ionization of their surface through electrostatic interaction with the cell membrane. The higher cleavage of the polymer at the endosomal pH dissociated the micellar structure and made the system very membrane-disruptive. This resulted in the enhancement of intracellular delivery via the endosomal pathway. Thus these results can be used for efficient tumor-specific uptake and intracellular delivery.

MSN are an important group of DGDS [144, 145]. The various MSN-based DDS possessing pH-responsive properties have been reviewed in the literature [146]. Due to their large surface area and considerable pore volumes, MSN can deliver a large quantity of cargo molecules under controlled conditions [147–149], and they are capable of releasing the cargo via stimuli such as pH [41, 150, 151]. The use of a pH-sensitive compartment means that they can be considered for use in pH-responsive delivery systems [152]. In a study by Chen *et al* in 2015, mechanized MSN (MMSN) with acid-sensitive properties were designed. Therein, adamantine (AD) was conjugated to the MSN surface through an acid-sensitive intermediate linker, and then DOX was loaded into the MSN as a model drug. The results showed that drug release at pH 7.4 was negligible, while at pH 5.5 efficient DOX release (up to 90%) was achieved. Thus, this pH-sensitive MSN displayed an accelerated drug release profile and enhanced inhibition of cellular proliferation in tumor cells [153]. In 2012, Gan *et al* [154] synthesized a system for the delivery of dexamethasone based on MSN with magnetic nanoparticle (Fe_3O_4) caps as nano-gates anchored in their pores. The pore openings were functionalized by chelating superparamagnetic Fe_3O_4 NP with an acid-labile 1,3,5-triaza adamantane (TAA) moiety. Rapid drug release was triggered by the endosomal pH (5.0–6.0) via pH-driven linkage disintegration; there was zero premature release in the physiological milieu (pH 7.4). The results showed that the nanocarriers entered into the cells via endocytosis and released the encapsulated cargos into the cytosol. Hence, such systems can be used as a nanogate to regulate the release rate and dosage of drug payloads.

2.4 Challenges and applications

2.4.1 Cancer therapy

In recent years, a large number of advances in smart DGDS based on pH responsive nanocarriers have been achieved, leading to significant breakthroughs in the therapy and diagnosis of various diseases and disorders, including cancers, pre-malignancies and infections. pH-responsive nanocarriers with the ability to encapsulate drugs and genes and release them in target tissues have been developed, thus avoiding the side-effects of the drugs on normal cells. In recent studies, issues such as the effects of pH-responsive nanocarriers (e.g. micelles and liposomes), the way that pH changes affect anticancer drug uptake by tumor and cancer cells, and the inhibition of tumor growth have been considered, as illustrated in figure 2.7.

Figure 2.8 illustrates drug release for liposomes via the dissolution of the nanocarrier with pH variation. The pH-sensitive liposomes release the loaded drug as a result of acidification in the endosomal compartment or pathological tissues such as cancers. In addition, because of the fusion between the liposomes and the endosomal membrane, the encapsulated drugs can be released directly into the cytoplasm [157].

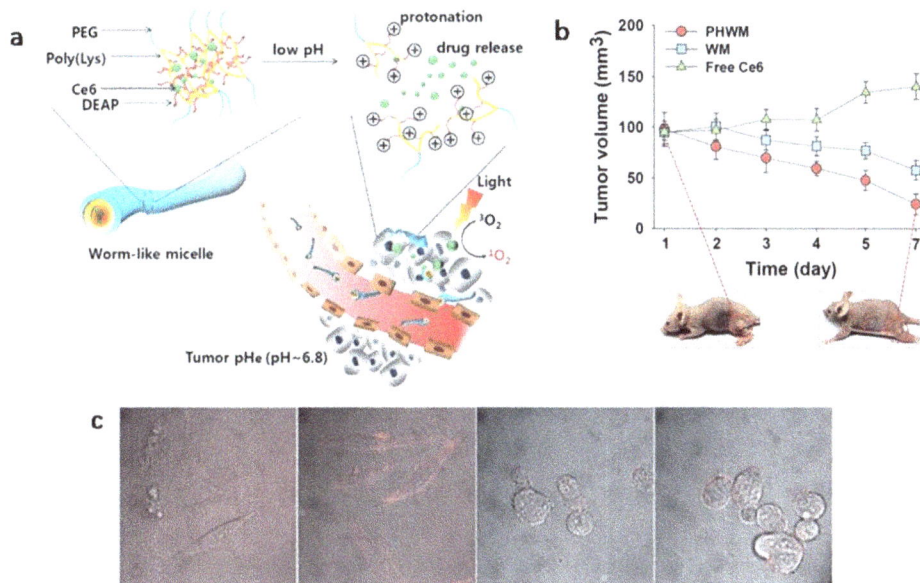

Figure 2.7. (a) Schematic of the structure and targeted delivery of mPEG-ser-[poly(Lys-DEAP)]$_2$-based worm-like micelles to tumor sites. (b) Effect of the nanocarriers and pH changes in cancer therapy; reduction in tumor volume caused by photodynamic tumor growth inhibition through the phototoxicity of injected pH-responsive worm-like micelles (PHWMs) disintegrated in the acidic milieu of solid tumors. (a), (b) Reproduced with permission from [152]. Copyright 2014 by The Royal Society of Chemistry. (c) Enhanced pH-dependent cellular uptake of nanogel clusters upon 30 min incubation through reduction of pH shown by confocal images (overlap of rhodamine emission and DIC image): from left to right, HeLa cell line at pH 7.4 and pH 6.5, and MCF-7 cell line at pH 7.4 and pH 6.5. Reproduced with permission from [153]. Copyright 2014 by The Royal Society of Chemistry.

2.4.2 Multi/dual-pH-responsive drug delivery systems

Multi-pH-responsive DDS enable us to develop new methods for high-efficiency therapies. For example, Du *et al* designed a pH-responsive polymer–doxorubicin conjugate delivery system that can respond to both extracellular and intracellular pH variations. They showed that the surface charge of this dual-pH-sensitive nanocarrier was reversed from positive to negative at the extracellular pH of tumors (~6.8), facilitating the cellular uptake. Also, in the more acidic pH conditions of the intracellular endosomal compartment (~5.0), a higher drug release rate was obtained from the endocytosed nanocarriers. Furthermore, the cytotoxicity of the delivery systems was intensified and, notably, the resistance of cancer cells to the anticancer drug was reduced [158].

Dual-pH-sensitive nanocarriers have been developed that are responsive to both tumor and endosomal pH. These were composed of a self-assembled copolymer coupled with two distinct groups of acid-labile linkers. Consequently, the nanocarrier was influenced by the different pH values in tumor tissues and endosomal compartments. Both cellular uptake of the anticancer drug-loaded nanocarrier and drug release were enhanced, resulting in more efficient cancer therapy [159].

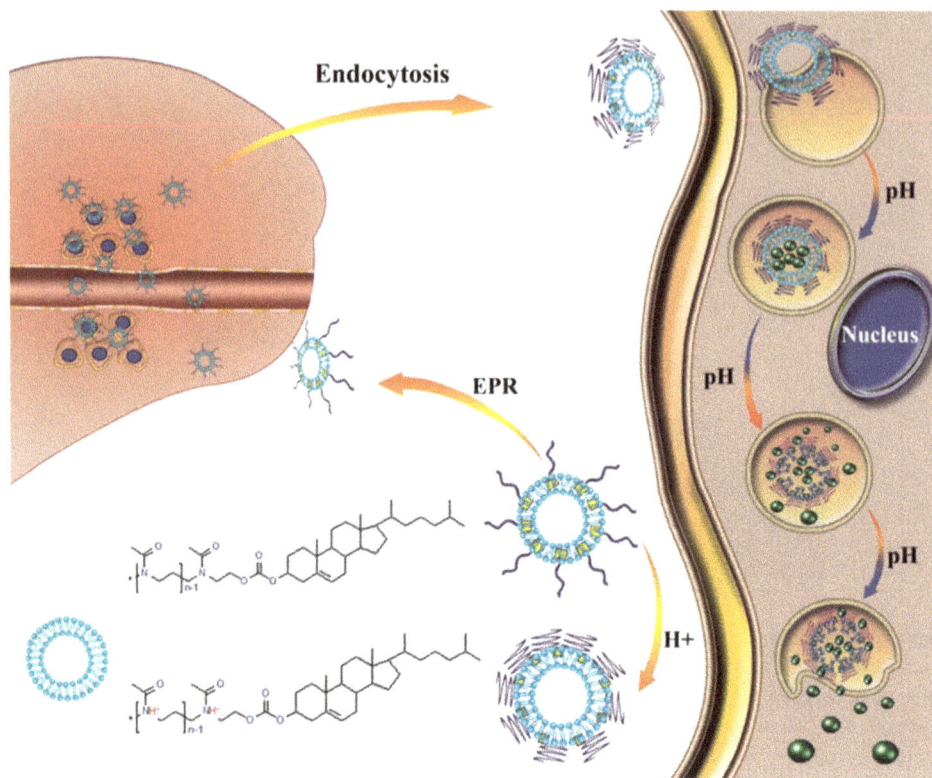

Figure 2.8. Schematic representation of the delivery of liposomes and subsequent drug release.

2.4.3 Simultaneous hydrophilic/hydrophobic oral delivery

In an attempt to design delivery systems capable of carrying both hydrophobic and hydrophilic molecular drugs, Watkins *et al* [160] developed an anionic pseudo-peptidic polymer as a hydrogel backbone with pH-sensitivity and amphiphilicity, crosslinking it with L-lysine methyl ester. Hydrophobic fluorescein and hydrophilic fluorescein-dextran were used as model drugs. It was shown that drug release for both model drugs was significantly higher at pH 7.4 than at pH 3.0. This DDA showed various advantages, including the ability to carry a broad size range of cargo molecules and macromolecules (e.g. proteins, antibodies, plasmids, viruses and NP) for targeted delivery, and responsiveness to pH alterations in the GIT. In addition, this pseudo-peptidic hydrogel facilitated the oral delivery of large payloads (e.g. larger than 70 kDa).

2.4.4 Remote-controlled pH-responsive smart delivery systems

In a recent investigation, a hydrazide-tailored PMMA-coated magnetic nanocube was developed that made remote control of pH-responsive DDS possible through external stimulation, e.g. a magnetic field. Such systems have also resulted in targeting to tumor tissue and efficient anticancer drug release with subsequent cancer cell apoptosis [161].

2.4.5 Highly efficient anticancer therapy

In recent years, progress in the development of anticancer DDS with efficient cancer treatment capability and good tumor targeting efficiency has been the subject of various studies. Such systems have been reported for the delivery of DOX anticancer drug, e.g. a malachite green carbinol base–DOX–liposome (MG–DOX–L) system [162] and a docetaxel (DTX)-loaded PEG–hyperbranched polyacylhydrazone (PEG–HPAH–DTX) system [163].

In several studies, obtaining a sufficiently deep penetration of the anticancer drug into the tumor tissue has been the main challenge. It was shown that by controlling several parameters, such as the surface charge and size of the nanoparticle carriers, responses to slight changes of pH can be optimized according to the acidic pH milieu of tumor tissues. Therefore, a deep penetration of anticancer nanocarriers into the tumor tissue followed by cellular uptake via endocytosis in cancer cells can be accomplished [156]. In 2015, Kang *et al* fabricated polymeric micelle nanocarriers with pH-sensitive hydrazone covalently bonded linkers via self-assembly of dendritic polyrotaxane drug–polymer conjugates. No premature burst release was detected at physiological pH, but it was possible to deliver the anticancer drug passively to the nuclei of the tumor cells (figure 2.9(a)) and release it efficiently at mildly acidic pH (6.0 and 5.0), leading to significant anticancer efficacy and cytotoxic effects compared to the parent DOX (figure 2.9(b)). Figure 2.9(c) presents fluorescence images of the intracellular uptake of DOX-conjugated micelle nanocarriers by A549 lung cancer cells incubated for 1, 4 and 8 h. However, the DOX-free micelle nanocarriers showed negligible toxicity and promising tolerability as carriers [164].

In another attempt, Wang *et al* synthesized pH-sensitive PEG–PCL-based micelle hydrogels (20 nm) loaded with DTX and investigated their suitability as oral delivery systems and their responsiveness to pH conditions in the GIT. According to the results, the cytotoxicity of the DTX-encapsulated micelles was higher than that of free DTX, based on the 4T1 cancer cell viability test (figure 2.10(a)). The nanocarriers also showed strong pH-dependent swelling behavior as the pH altered from 2 to 8 (figure 2.10(b)). Furthermore, the rapid diffusion of the DTX–micelle–hydrogel nanocarriers in simulated intestinal fluid (in the presence of trypsin) with a high pH value (i.e. 6.8) by comparison to simulated gastric fluid (in the presence of pepsin) with a low pH value (i.e. 1.2) was reported (figure 2.10(c) and (d)). Such pH changes suggested that targeted delivery of DTX to the small intestine might be possible. Furthermore, the DTX-loaded micelle–hydrogel nanocarriers enhanced the oral bioavailability of DTX (up to 75.6%) by about tenfold over DTX micelles without hydrogel, with this resulting from the targeted-release ability of the pH-sensitive hydrogel. Efficient inhibition of tumor growth in a subcutaneous 4T1 breast cancer model, together with reduced systemic toxicity, was shown [165].

2.4.6 Developing innovative pH-tunable nanocarriers

Generating innovative pH-responsive particles and nanosystems with advanced features is of great importance. For example, a new pH-sensitive particle called the amphiphilic Janus particle was developed. This comprised emulsion droplets capable

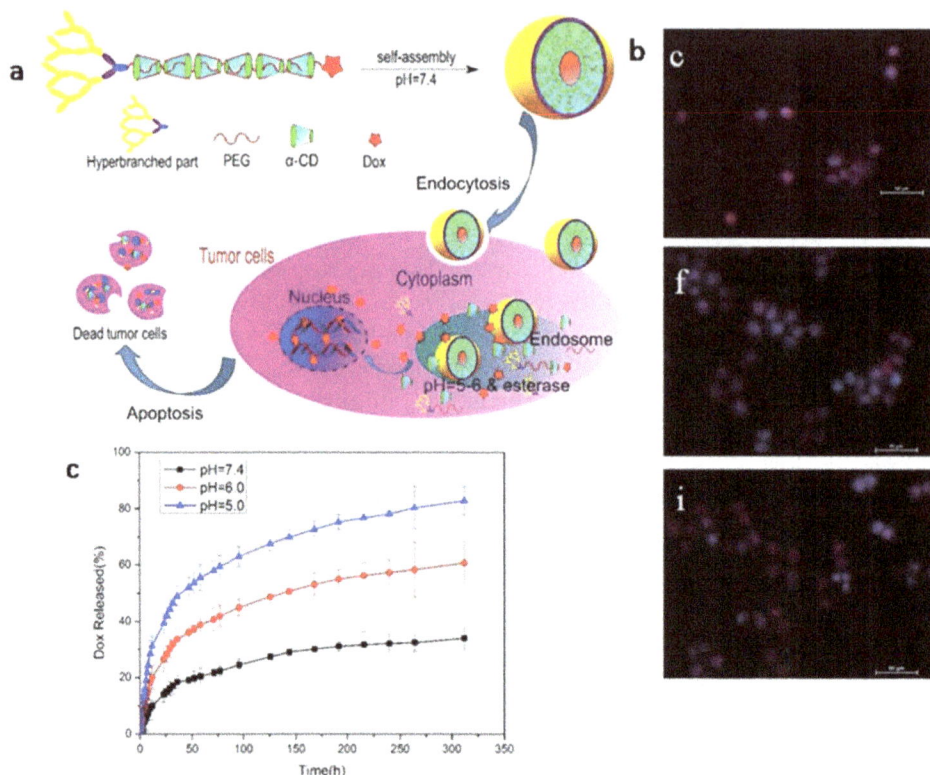

Figure 2.9. (a) Preparation process for dendritic polyrotaxane drug–polymer conjugates forming micelles via self-assembly followed by selective drug release within the endosomal compartments of tumor cells. (b) The overlays of two fluorescence imagings of intracellular uptake by A549 lung cancer cells incubated with DOX-conjugated micelles (PR–g–DOX micelles) for 1, 4 and 8 h (upper, middle and lower images, respectively); employing DAPI (blue fluorescence) for cell nuclei staining, and DOX for fluorescence in cells. (c) pH-responsive hydrolysis of dendritic polyrotaxane drug conjugate nanocarriers with hydrazone bonds in acidic conditions (endosomes or tumors), leading to fast DOX release. Reproduced with permission from [164]. Copyright 2015 by The Royal Society of Chemistry.

of stabilizing multiphase fluidic compounds and possessing dynamically tunable shape properties, so that its morphology or assembled shape changes drastically in response to alterations of the surfactant properties induced by phase inversion emulsification triggered by pH changes in the medium (e.g. aqueous solution). One side of the particles was concentrated with hydrophobic monomers (e.g. styrene) and the other side was concentrated with pH-responsive hydrophilic units (e.g. acrylic acid). Thereafter, the pH-induced aggregation/dispersion behavior of such particles could be changed, leading to shape transformation with no induced destabilization. Such systems could supply new designs for pH-responsive DGDS [166].

2.4.7 Dual cargo pH-responsive delivery systems

pH-responsive systems have enabled the development of controlled delivery and release for dual cargo-containing nanocarriers. Therein, each cargo can be released

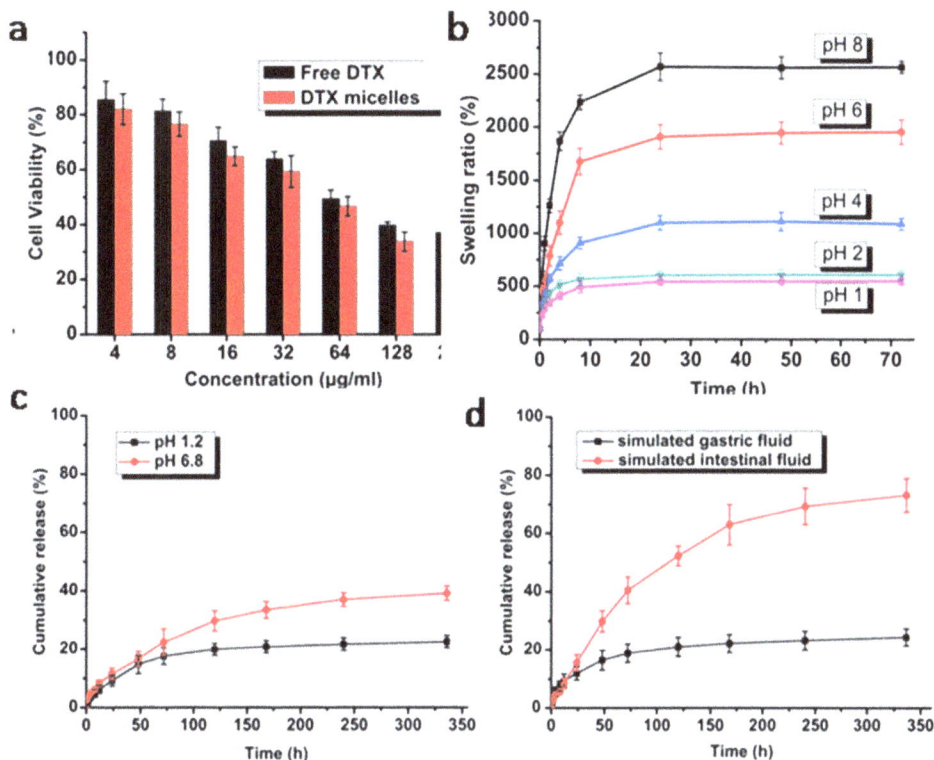

Figure 2.10. (a) Cell viability of free DTX and DTX-loaded micelle–hydrogel nanocarriers; (b) the micelle–hydrogel swelling ratio response to pH changes; (c), (d) the cumulative release profile of DTX-loaded nanocarriers with respect to pH changes and simulated gastric and intestinal environments, respectively. Reproduced with permission from [162]. Copyright 2014 Elsevier.

separately at a distinct pH value, as opposed to the release of one component at a higher pH value and the other at a lower one. This makes the controlled and accurate delivery of several different cargos possible [167]. A crucial application of such systems is the design of anticancer therapeutic systems to overcome multi-drug resistance (MDR) against the anticancer drug. MDR is a significant cause of failed clinical cancer treatment. A novel method to design pH-responsive DDS [168] containing combinational delivery systems was reported [134]. The efficient utilization of these pH-responsive DDS in the co-delivery of small interfering RNA (siRNA) and anticancer drugs has been reported. For example, in an effort to eliminate the MDR caused by cancer cells, a delivery system was designed by Yu *et al.* A nucleic acid (siRNA) was administered in the pH-responsive delivery system to induce efficient silencing of gene expressions (Luc expression) in the carcinoma cell lines. In addition, the anticancer drug paclitaxel (PTX) was simultaneously loaded into the DDS as the chemotherapeutic agent [169].

2.4.8 Targeting of the body compartments via pH-responsive DGDS

The pH value alterations in various biological compartments of the body, such as organs, tissues and cells present in the GIT (e.g. the ileum and small intestine), as well as the endocytic compartments in cells, can be considered in the design of pH-responsive DDS [170]. For example, Fu *et al* synthesized a chitosan-modified gold nanoparticle attached onto the negatively charged surface of liposomes. This nanocarrier was used for the delivery of a gastric antimicrobial drug (doxycycline) to treat a stomach infection caused by *Helicobacter pylori* bacteria. Although this DDS showed stability at gastric pH, it became unstable at neutral pH via the detachment of gold NP, the liposomes actively fused within the bacterial membrane and thereafter drug release occurred [171]. Gogoi *et al* showed that a DDS based on carbon dot-coated calcium alginate bead (CA–CD) nanocarriers has a good drug release rate at the low pH of organs such as the GIT [172].

2.4.9 Improved drug/gene DDS parameters

The modification of parameters influencing the efficiency of DDS, including duration in the blood circulation, bioavailability, solubility, biodistribution and the biodegradability of the carriers in the biological milieu after delivery of the cargo, should all be considered. There are also several limitations to these smart delivery systems, such as poor oral availability, unpredictable drug release rate, poor stability, an insufficient level of sensitivity to small changes in pH values and possible damage to the cargo caused by the nanocarriers themselves, and these should be eliminated [2, 173]. Wang *et al* reported that the oral bioavailability of anticancer drugs can be enhanced by pH-responsive DDS [165]. pH-responsive DDS have also been shown to eliminate the premature degradation of the drugs in lysosomes. In this manner, endocytosis and the subsequent fusion of the DDS via rupture or destabilization of the endosomal membrane (due to lower endosomal pH) can be obtained. The drug release rate in the cytoplasm is then enhanced [174].

pH-responsive DDS can potentially be used to improve parameters such as drug dosage, release rate and loading capacity. Higher drug-loading capacity through use of polymer/silica-based nanocarriers [4] and PEG–PCL-based micelle hydrogels [165] has been reported. Importantly, where there is limited bioavailability for drugs, as with therapeutic proteins with a high isoelectric point within the small intestine and the intestinal epithelium, the need to provide a sufficient quantity of therapeutic cargo through a single administration (e.g. via the oral route) should be considered. To achieve this goal, the time-limited response to pH-alterations can be modified [175]. In an attempt to obtain long-term drug availability, pH CA–CD nanocarriers were synthesized and exhibited good stability and swelling properties [172].

2.4.10 Biomolecule delivery via pH-sensitive nanocarriers

Several attempts have been made to develop systems for the delivery of proteins and peptide-molecule drugs. Choi *et al* designed an insulin pH-sensitive delivery system that overcame drawbacks such as the crowding effect and biomolecular denaturation. To this end, dendritic polyglycerol (dPG) nanogels with acid-labile benzacetal linkers were prepared. In drug delivery analysis, they reported the elimination of protein activity loss during delivery in a mild pH environment [176, 177]. Furthermore, another important challenge relates to the shortage or accumulation of metal ions in biological environments. Alterations in the concentration of metal ions can induce disorders such as Alzheimer's disease (due to the dysregulation of copper ions and the subsequent aggregation of amyloid-beta (Aβ) peptides in the brain) [178]. Thus the response of such disorders to pH-alterations has been reported [179]. Ejima *et al* proposed that metal ion cargos could be carried and released in cells through the administration of pH-triggered delivery systems [180]. Accordingly, such pH-responsive DDS could potentially be applied in the treatment of disorders that show pH-dependence.

2.4.11 Combination with photodynamic therapy

The combination of drug delivery and photodynamic therapy (PDT) systems is of great importance. For example, fullerenes (i.e. closed carbon nanocages) have been employed in such systems. Shi *et al* developed a DOX delivery system that uses poly(ethylenimine) derivatized fullerene (DOX–EI–C60) as the nanocarrier. Here, the DOX is covalently conjugated on PEI–C60 nanocarrier. It was shown that the drug release was strongly dependent on the pH value of the acidic environment of the tumor, using *in vivo* and *in vitro* studies. Furthermore, several advantageous outcomes were achieved, such as strong cytotoxicity to tumor cells, effective tumor targeting and efficient tumor growth inhibition. This system showed a synergistic effect on both PDT and chemotherapy and, in addition, no systemic toxicity toward normal cells was reported [181].

References

[1] Joshi G K, Johnson M A and Sardar R 2014 Novel pH-responsive nanoplasmonic sensor: controlling polymer structural change to modulate localized surface plasmon resonance response *RSC Adv.* **4** 15807–15

[2] Yu B, Li X, Zheng W, Feng Y, Wong Y-S and Chen T 2014 pH-responsive cancer-targeted selenium nanoparticles: a transformable drug carrier with enhanced theranostic effects *J. Mater. Chem.* B **2** 5409–18

[3] He L *et al* 2014 Carbon nanodots@zeolitic imidazolate framework-8 nanoparticles for simultaneous pH-responsive drug delivery and fluorescence imaging *CrystEngComm.* **16** 3259–63

[4] Rasouli S, Davaran S, Rasouli F, Mahkam M and Salehi R 2014 Synthesis, characterization and pH-controllable methotrexate release from biocompatible polymer/silica nanocomposite for anticancer drug delivery *Drug Deliv.* **21** 155–63

[5] Alvarez-Lorenzo C and Concheiro A 2014 Smart drug delivery systems: from fundamentals to the clinic *Chem. Commun.* **50** 7743–65

[6] Alvarez-Lorenzo C and Concheiro A 2013 From drug dosage forms to intelligent drug-delivery systems: a change of paradigm *Smart Mater. Drug Deliv.* **1** 1

[7] Wike-Hooley J, Haveman J and Reinhold H 1984 The relevance of tumour pH to the treatment of malignant disease *Radiother. Oncol.* **2** 343–66

[8] Vaupel P, Kallinowski F and Okunieff P 1989 Blood flow, oxygen and nutrient supply and metabolic microenvironment of human tumors: a review *Cancer Res.* **49** 6449–65

[9] Tian L and Bae Y H Cancer nanomedicines targeting tumor extracellular pH *Colloid. Surface.* B **99** 116–26

[10] Hsu P P and Sabatini D M 2008 Cancer cell metabolism: Warburg and beyond *Cell* **134** 703–7

[11] Vander Heiden M G, Cantley L C and Thompson C B 2009 Understanding the Warburg effect: the metabolic requirements of cell proliferation *Science* **324** 1029–33

[12] Volk T, Jähde E, Fortmeyer H, Glüsenkamp K and Rajewsky M 1993 pH in human tumour xenografts: effect of intravenous administration of glucose *Br. J. Cancer* **68** 492

[13] Mellman I, Fuchs R and Helenius A 1986 Acidification of the endocytic and exocytic pathways *Annu. Rev. Biochem.* **55** 663–700

[14] Bae Y, Nishiyama N, Fukushima S, Koyama H, Yasuhiro M and Kataoka K 2005 Preparation and biological characterization of polymeric micelle drug carriers with intracellular pH-triggered drug release property: tumor permeability, controlled subcellular drug distribution and enhanced *in vivo* antitumor efficacy *Bioconjugate Chem.* **16** 122–30

[15] Liu J *et al* 2013 pH-sensitive nano-systems for drug delivery in cancer therapy *Biotechnol. Adv.* **32** 693–710

[16] Lv Y, Huang H, Yang B, Liu H, Li Y and Wang J 2014 A robust pH-sensitive drug carrier: aqueous micelles mineralized by calcium phosphate based on chitosan *Carbohyd. Polym.* **111** 101–7

[17] Quader S *et al* 2014 Selective intracellular delivery of proteasome inhibitors through pH-sensitive polymeric micelles directed to efficient antitumor therapy *J. Controlled Release* **188** 67–77

[18] Kamimura M and Nagasaki Y 2013 pH-sensitive polymeric micelles for enhanced intracellular anti-cancer drug delivery *J. Photopolym. Sci. Technol.* **26** 161–4

[19] Wu H, Zhu L and Torchilin V P 2013 pH-sensitive poly (histidine)–PEG/DSPE–PEG co-polymer micelles for cytosolic drug delivery *Biomaterials* **34** 1213–22

[20] Kataoka K, Harada A and Nagasaki Y 2001 Block copolymer micelles for drug delivery: design, characterization and biological significance *Adv. Drug Deliv. Rev.* **47** 113–31

[21] Hatakeyama H *et al* 2014 The systemic administration of an anti-miRNA oligonucleotide encapsulated pH-sensitive liposome results in reduced level of hepatic microRNA-122 in mice *J. Controlled Release* **173** 43–50

[22] Ramasamy T *et al* 2014 Layer-by-layer assembly of liposomal nanoparticles with PEGylated polyelectrolytes enhances systemic delivery of multiple anticancer drugs *Acta Biomater.* **10** 5116–27

[23] Yoshizaki Y, Yuba E, Sakaguchi N, Koiwai K, Harada A and Kono K 2014 Potentiation of pH-sensitive polymer-modified liposomes with cationic lipid inclusion as antigen delivery carriers for cancer immunotherapy *Biomaterials* **35** 8186–96

[24] Watarai S, Iwase T, Tajima T, Yuba E and Kono K 2013 Efficiency of pH-sensitive fusogenic polymer-modified liposomes as a vaccine carrier *Sci. World J.* **2013**

[25] Xu H *et al* 2014 The bifunctional liposomes constructed by poly (2-ethyl-oxazoline)-cholesteryl methyl carbonate: an effectual approach to enhance liposomal circulation time, pH-sensitivity and endosomal escape *Pharmaceut. Res.* **31** 3038–50

[26] Torchilin V P 2005 Recent advances with liposomes as pharmaceutical carriers *Nat. Rev. Drug Discovery* **4** 145–60

[27] Torchilin V P 2007 Micellar nanocarriers: pharmaceutical perspectives *Pharmaceuti. Res.* **24** 1–16

[28] Sihorkar V and Vyas S 2001 Potential of polysaccharide anchored liposomes in drug delivery, targeting and immunization *J. Pharm. Pharm. Sci.* **4** 138–58

[29] Raemdonck K, Demeester J and De Smedt S 2009 Advanced nanogel engineering for drug delivery *Soft Matter* **5** 707–15

[30] Soni G and Yadav K S 2014 Nanogels as potential nanomedicine carrier for treatment of cancer: a mini review of the state of the art *Saudi Pharmaceut. J.* **2014**

[31] Abandansari H S, Nabid M R, Rezaei S J T and Niknejad H 2014 pH-sensitive nanogels based on Boltorn® H40 and poly (vinylpyridine) using mini-emulsion polymerization for delivery of hydrophobic anticancer drugs *Polymer* **55** 3579–90

[32] Chacko R T, Ventura J, Zhuang J and Thayumanavan S 2012 Polymer nanogels: a versatile nanoscopic drug delivery platform *Adv. Drug Deliv. Rev.* **64** 836–51

[33] Du C *et al* 2013 A pH-sensitive doxorubicin prodrug based on folate-conjugated BSA for tumor-targeted drug delivery *Biomaterials* **34** 3087–97

[34] Lv S *et al* 2014 Well-defined polymer–drug conjugate engineered with redox and pH-sensitive release mechanism for efficient delivery of paclitaxel *J. Controlled Release* **194** 220–7

[35] Rigogliuso S, Sabatino M A, Adamo G, Grimaldi N, Dispenza C and Ghersi G 2012 Polymeric nanogels: nanocarriers for drug delivery application *Chem. Eng.* **2012** 27

[36] Xu D, Wu F, Chen Y, Wei L and Yuan W 2013 pH-sensitive degradable nanoparticles for highly efficient intracellular delivery of exogenous protein *Int. J. Nanomed.* **8** 3405

[37] Soppimath K S, Aminabhavi T M, Kulkarni A R and Rudzinski W E 2001 Biodegradable polymeric nanoparticles as drug delivery devices *J. Controlled Release* **70** 1–20

[38] Liu R *et al* 2011 Anti-tumor drug delivery of pH-sensitive poly (ethylene glycol)-poly (L-histidine-)-poly (L-lactide) nanoparticles *J. Controlled Release* **152** 49–56

[39] Ma L, Liu M, Liu H, Chen J and Cui D 2010 *In vitro* cytotoxicity and drug release properties of pH- and temperature-sensitive core–shell hydrogel microspheres *Int. J. Pharmaceut.* **385** 86–91

[40] Mukhopadhyay P, Chakraborty S, Bhattacharya S, Mishra R and Kundu P 2015 pH-sensitive chitosan/alginate core–shell nanoparticles for efficient and safe oral insulin delivery *Int. J. Biol. Macromol.* **72** 640–8

[41] Yuan L, Tang Q, Yang D, Zhang J Z, Zhang F and Hu J 2011 Preparation of pH-responsive mesoporous silica nanoparticles and their application in controlled drug delivery *J. Phys. Chem.* C **115** 9926–32

[42] Zheng Q *et al* 2014 Mussel-inspired polydopamine coated mesoporous silica nanoparticles as pH-sensitive nanocarriers for controlled release *Int. J. Pharmaceut.* **463** 22–6

[43] Yang S *et al* 2012 A facile preparation of targetable pH-sensitive polymeric nanocarriers with encapsulated magnetic nanoparticles for controlled drug release *J. Mater. Chem.* **22** 25354–61

[44] Kim S and Park C B 2010 Mussel-inspired transformation of $CaCO_3$ to bone minerals *Biomaterials* **31** 6628–34

[45] Wei W *et al* 2008 Preparation of hierarchical hollow $CaCO_3$ particles and the application as anticancer drug carrier *J. Am. Chem. Soc.* **130** 15808–10

[46] Deng Z, Zhen Z, Hu X, Wu S, Xu Z and Chu P K 2011 Hollow chitosan–silica nanospheres as pH-sensitive targeted delivery carriers in breast cancer therapy *Biomaterials* **32** 4976–86

[47] Sonaje K *et al* 2010 Self-assembled pH-sensitive nanoparticles: a platform for oral delivery of protein drugs *Adv. Funct. Mater.* **20** 3695–700

[48] Boppana R, Mohan G K, Nayak U, Mutalik S, Sa B and Kulkarni R V 2015 Novel pH-sensitive IPNs of polyacrylamide-g-gum ghatti and sodium alginate for gastro-protective drug delivery *Int. J. Biol. Macroml.* **75** 133–43

[49] Rao K M, Rao K K, Ramanjaneyulu G and Ha C-S 2015 Curcumin encapsulated pH sensitive gelatin based interpenetrating polymeric network nanogels for anti cancer drug delivery *Int. J. Pharmaceut.* **478** 788–95

[50] Zhang H, Wu H, Fan L, Li F, Ch Gu and Jia M 2009 Preparation and characteristics of pH-sensitive derivated dextran hydrogel nanoparticles *Polym. Compos.* **30** 1243–50

[51] Wang X-Q and Zhang Q 2012 pH-sensitive polymeric nanoparticles to improve oral bioavailability of peptide/protein drugs and poorly water-soluble drugs *Eur. J. Pharmaceut. Biopharmaceut.* **82** 219–29

[52] Kamada H *et al* 2004 Design of a pH-sensitive polymeric carrier for drug release and its application in cancer therapy *Clin. Cancer Res.* **10** 2545–50

[53] Na K, Lee E S and Bae Y H 2003 Adriamycin loaded pullulan acetate/sulfonamide conjugate nanoparticles responding to tumor pH: pH-dependent cell interaction, internalization and cytotoxicity *in vitro J. Controlled Release* **87** 3–13

[54] Stayton P *et al* 2005 'Smart' delivery systems for biomolecular therapeutics *Orthodont. Cran. Res.* **8** 219–25

[55] Ge Y, Li S, Wang S and Moore R 2014 *Nanomedicine: Principles and Perspectives* (Berlin: Springer)

[56] Gillies E R, Jonsson T B and Fréchet J M 2004 Stimuli-responsive supramolecular assemblies of linear-dendritic copolymers *J. Am. Chem. Soc.* **126** 11936–43

[57] Ganta S, Talekar M, Singh A, Coleman T P and Amiji M M 2014 Nanoemulsions in translational research—opportunities and challenges in targeted cancer therapy *AAPS PharmSciTech.* **15** 694–708

[58] Dai S, Ravi P and Tam K C 2008 pH-responsive polymers: synthesis, properties and applications *Soft Matter* **4** 435–49

[59] Jones R *et al* 2003 Poly (2-alkylacrylic acid) polymers deliver molecules to the cytosol by pH-sensitive disruption of endosomal vesicles *Biochem J.* **372** 65–75

[60] Thomas J L, Barton S W and Tirrell D A 1994 Membrane solubilization by a hydrophobic polyelectrolyte: surface activity and membrane binding *Biophys. J.* **67** 1101–6

[61] Thomas J L and Tirrell D A 1992 Polyelectrolyte-sensitized phospholipid vesicles *Acc. Chem. Res.* **25** 336–42

[62] Foster S, Duvall C L, Crownover E F, Hoffman A S and Stayton P S 2010 Intracellular delivery of a protein antigen with an endosomal-releasing polymer enhances CD8 T-cell production and prophylactic vaccine efficacy *Bioconjugate Chem.* **21** 2205–12

[63] Ding Y, Gao J, Yang X, He J, Zhou Z and Hu Y 2014 Preparation of water dispersible, fluorescent Ag–PAA–PVP hybrid nanogels and their optical properties *Adv. Powder Technol.* **25** 244–9

[64] Park S Y and Bae Y H 1999 Novel pH-sensitive polymers containing sulfonamide groups *Macromol. Rapid Commun.* **20** 269–73

[65] Murthy N, Campbell J, Fausto N, Hoffman A S and Stayton P S 2003 Bioinspired pH-responsive polymers for the intracellular delivery of biomolecular drugs *Bioconjugate Chem.* **14** 412–9

[66] Poon G and Gariepy J 2007 Cell-surface proteoglycans as molecular portals for cationic peptide and polymer entry into cells *Biochem. Soc. Trans.* **35** 788–93

[67] Erbacher P, Remy J and Behr J 1999 Gene transfer with synthetic virus-like particles via the integrin-mediated endocytosis pathway *Gene Ther.* **6** 1

[68] Boussif O *et al* 1995 A versatile vector for gene and oligonucleotide transfer into cells in culture and *in vivo*: polyethylenimine *Proc. Natl Acad. Sci.* **92** 7297–301

[69] Pack D W, Putnam D and Langer R 2000 Design of imidazole-containing endosomolytic biopolymers for gene delivery *Biotechnol. Bioeng.* **67** 217–23

[70] Pietersz G A, Tang C-K and Apostolopoulos V 2006 Structure and design of polycationic carriers for gene delivery *Mini Rev. Med. Chem.* **6** 1285–98

[71] Kitazoe M, Futami J, Nishikawa M, Yamada H and Maeda Y 2010 Polyethylenimine-cationized β-catenin protein transduction activates the Wnt canonical signaling pathway more effectively than cationic lipid-based transduction *Biotechnol. J.* **5** 385–92

[72] Futami J *et al* 2005 Intracellular delivery of proteins into mammalian living cells by polyethylenimine-cationization *J. Biosci. Bioeng.* **99** 95–103

[73] Zauner W, Ogris M and Wagner E 1998 Polylysine-based transfection systems utilizing receptor-mediated delivery *Adv. Drug Deliv. Rev.* **30** 97–113

[74] Boussif O, Zanta M A and Behr J-P 1996 Optimized galenics improve *in vitro* gene transfer with cationic molecules up to 1000-fold *Gene Ther.* **3** 1074–80

[75] Kichler A, Leborgne C, Coeytaux E and Danos O 2001 Polyethylenimine-mediated gene delivery: a mechanistic study *J. Gene Med.* **3** 135–44

[76] Oh Y, Suh D, Kim J, Choi H, Shin K and Ko J 2002 Polyethylenimine-mediated cellular uptake, nucleus trafficking and expression of cytokine plasmid DNA *Gene Ther.* **9** 23

[77] Segura T and Hubbell J A 2007 Synthesis and *in vitro* characterization of an ABC triblock copolymer for siRNA delivery *Bioconjugate Chem.* **18** 736–45

[78] Park S-C, Nam J-P, Kim Y-M, Kim J-H, Nah J-W and Jang M-K 2013 Branched polyethylenimine-grafted-carboxymethyl chitosan copolymer enhances the delivery of pDNA or siRNA *in vitro* and *in vivo* *International Journal of Nanomedicine* **8** 3663

[79] Meng F, Zhong Y, Cheng R, Deng C and Zhong Z 2014 pH-sensitive polymeric nanoparticles for tumor-targeting doxorubicin delivery: concept and recent advances *Nanomedicine* **9** 487–99

[80] Chen M-C *et al* 2013 Recent advances in chitosan-based nanoparticles for oral delivery of macromolecules *Adv. Drug Deliv. Rev.* **65** 865–79

[81] Bugamelli F, Raggi M A, Orienti I and Zecchi V 1998 Controlled insulin release from chitosan microparticles *Arch. Pharm.* **331** 133–8

[82] Jayakumar R, Reis R and Mano J 2006 Phosphorous containing chitosan beads for controlled oral drug delivery *J. Bioact. Compat. Pol.* **21** 327–40

[83] Karimi M, Avci P, Mobasseri R, Hamblin M R and Naderi-Manesh H 2013 The novel albumin–chitosan core–shell nanoparticles for gene delivery: preparation, optimization and cell uptake investigation *J. Nanoparticle Res.* **15** 1–14

[84] Karimi M, Avci P, Ahi M, Gazori T, Hamblin M R and Naderi-Manesh H 2013 Evaluation of chitosan-tripolyphosphate nanoparticles as a p-shRNA delivery vector: formulation, optimization and cellular uptake study *J. Nanopharmaceut. Drug Delivery* **1** 266–78

[85] Yang J, Chen J, Pan D, Wan Y and Wang Z 2013 pH-sensitive interpenetrating network hydrogels based on chitosan derivatives and alginate for oral drug delivery *Carbohy. Polym.* **92** 719–25

[86] Jana S, Maji N, Nayak A K, Sen K K and Basu S K 2013 Development of chitosan-based nanoparticles through inter-polymeric complexation for oral drug delivery *Carbohy. Polym.* **98** 870–6

[87] Čalija B, Cekić N, Savić S, Daniels R, Marković B and Milić J 2013 pH-sensitive microparticles for oral drug delivery based on alginate/oligochitosan/Eudragit® L100-55 'sandwich' polyelectrolyte complex *Colloid. Surface.* B **110** 395–402

[88] Jahromi M A M, Karimi M, Azadmanesh K, Manesh H N, Hassan Z M and Moazzeni S M 2014 The effect of chitosan-tripolyphosphate nanoparticles on maturation and function of dendritic cells *Comparative Clin. Path.* **23** 1421–7

[89] Yang Y, Wang S, Wang Y, Wang X, Wang Q and Chen M 2014 Advances in self-assembled chitosan nanomaterials for drug delivery *Biotechnol. Adv.* **32** 1301–16

[90] Nogueira D R, Tavano L, Mitjans M, Pérez L, Infante M R and Vinardell M P 2013 *In vitro* antitumor activity of methotrexate via pH-sensitive chitosan nanoparticles *Biomaterials* **34** 2758–72

[91] Cavallaro G *et al* 2014 Synthesis and characterization of polyaspartamide copolymers obtained by ATRP for nucleic acid delivery *Int. J. Pharmaceut.* **466** 246–57

[92] Wang J *et al* 2014 Preparation and pH controlled release of polyelectrolyte complex of poly (l-malic acid-co-d, l-lactic acid) and chitosan *Colloid. Surface.* B **115** 275–9

[93] Lanz-Landázuri A, Martínez de Ilarduya A, García-Alvarez M and Muñoz-Guerra S 2014 Poly (β, L-malic acid)/Doxorubicin ionic complex: a pH-dependent delivery system *React. Funct. Polym.* **81** 45–53

[94] Wang M *et al* 2013 A pH-sensitive gene delivery system based on folic acid–PEG-chitosan–PAMAM–plasmid DNA complexes for cancer cell targeting *Biomaterials* **34** 10120–32

[95] Wen H, Guo J, Chang B and Yang W 2013 pH-responsive composite microspheres based on magnetic mesoporous silica nanoparticle for drug delivery *Eur. J. Pharmaceut. Biopharmaceut.* **84** 91–8

[96] Leung KC-F, Nguyen TD, Stoddart J F and Zink J I 2006 Supramolecular nanovalves controlled by proton abstraction and competitive binding *Chem. Mater.* **18** 5919–28

[97] Nguyen T D, Leung KC-F, Liong M, Pentecost C D, Stoddart J F and Zink J I 2006 Construction of a pH-driven supramolecular nanovalve *Org. Lett.* **8** 3363–6

[98] Roik N and Belyakova L 2014 Chemical design of pH-sensitive nanovalves on the outer surface of mesoporous silicas for controlled storage and release of aromatic amino acid *J. Solid State Chem.* **215** 284–91

[99] Tao C-A *et al* 2012 Fabrication of pH-sensitive graphene oxide–drug supramolecular hydrogels as controlled release systems *J. Mater. Chem.* **22** 24856–61

[100] Kavitha T, Abdi S I H and Park S-Y 2013 pH-sensitive nanocargo based on smart polymer functionalized graphene oxide for site-specific drug delivery *Phys. Chem. Chem. Phys.* **15** 5176–85

[101] Ren L, Liu T, Guo J, Guo S, Wang X and Wang W 2010 A smart pH responsive graphene/polyacrylamide complex via noncovalent interaction *Nanotechnology* **21** 335701

[102] Feng L, Li K, Shi X, Gao M, Liu J and Liu Z 2014 Smart pH-responsive nanocarriers based on nano-graphene oxide for combined chemo-and photothermal therapy overcoming drug resistance *Adv. Healthc. Mater.* **3** 1261–1

[103] Oupický D, Koňák Č, Ulbrich K, Wolfert M and Seymour L 2000 DNA delivery systems based on complexes of DNA with synthetic polycations and their copolymers *J. Controlled Release* **65** 149–1

[104] Greber U F, Willetts M, Webster P and Helenius A 1993 Stepwise dismantling of adenovirus 2 during entry into cells *Cell* **75** 477–86

[105] Rosenberg S A *et al* 1990 Gene transfer into humans—immunotherapy of patients with advanced melanoma, using tumor-infiltrating lymphocytes modified by retroviral gene transduction *New Engl. J. Med.* **323** 570–8

[106] Bett A J, Haddara W, Prevec L and Graham F L 1994 An efficient and flexible system for construction of adenovirus vectors with insertions or deletions in early regions 1 and 3 *Proc. Natl Acad. Sci.* **91** 8802–6

[107] Bramson J L, Graham F L and Gauldie J 1995 The use of adenoviral vectors for gene therapy and gene transfer *in vivo Curr. Opin. Biotechnol.* **6** 590–5

[108] Daya S and Berns K I 2008 Gene therapy using adeno-associated virus vectors *Clin. Microbiol. Rev.* **21** 583–93

[109] Wang A Y, Peng P D, Ehrhardt A, Storm T A and Kay M A 2004 Comparison of adenoviral and adeno-associated viral vectors for pancreatic gene delivery *in vivo Hum. Gene Ther.* **15** 405–13

[110] Walther W and Stein U 2000 Viral vectors for gene transfer *Drugs* **60** 249–1

[111] Cavazzana-Calvo M *et al* 2000 Gene therapy of human severe combined immunodeficiency (SCID)-X1 disease *Science* **288** 5466

[112] Hacein-Bey-Abina S *et al* 2002 Sustained correction of X-linked severe combined immunodeficiency by *ex vivo* gene therapy *New Engl. J. Med.* **346** 1185–93

[113] Hacein-Bey-Abina S *et al* 2010 Efficacy of gene therapy for X-linked severe combined immunodeficiency *New Engl. J. Med.* **363** 355–64

[114] Subbarao N K, Parente R A, Szoka F C Jr, Nadasdi L and Pongracz K 1987 The pH-dependent bilayer destabilization by an amphipathic peptide *Biochemistry* **26** 2964–72

[115] Kim A and Szoka F C Jr 1992 Amino acid side-chain contributions to free energy of transfer of tripeptides from water to octanol *Pharmaceut. Res.* **9** 504–14

[116] Sakurai Y *et al* 2011 Endosomal escape and the knockdown efficiency of liposomal-siRNA by the fusogenic peptide shGALA *Biomaterials* **32** 5733–42

[117] Wyman T B, Nicol F, Zelphati O, Scaria P, Plank C and Szoka F C 1997 Design, synthesis and characterization of a cationic peptide that binds to nucleic acids and permeabilizes bilayers *Biochemistry* **36** 3008–17

[118] Guo X D *et al* 2012 Oligomerized alpha-helical KALA peptides with pendant arms bearing cell-adhesion, DNA-binding and endosome-buffering domains as efficient gene transfection vectors *Biomaterials* **33** 6284–91

[119] McCarthy H O *et al* 2014 Development and characterization of self-assembling nanoparticles using a bio-inspired amphipathic peptide for gene delivery *J. Controlled Release* **189** 141–9

[120] Siepmann J, Siegel R A and Rathbone M J 2012 *Fundamentals and Applications of Controlled Release Drug Delivery* (Berlin: Springer)

[121] Liu Y, Wang W, Yang J, Zhou C and Sun J 2013 pH-sensitive polymeric micelles triggered drug release for extracellular and intracellular drug targeting delivery *Asian J. Pharmaceut. Sci.* **8** 159–67

[122] Heller J, Penhale D and Helwing R 1980 Preparation of polyacetals by the reaction of divinyl ethers and polyols *J. Polym. Sci. B Polym. Lett. Ed.* **18** 293–7

[123] Tu C *et al* 2013 Facile PEGylation of Boltorn® H40 for pH-responsive drug carriers *Polymer* **54** 2020–7

[124] Shenoi R A, Lai B F, Imran ul-haq M, Brooks D E and Kizhakkedathu J N 2013 Biodegradable polyglycerols with randomly distributed ketal groups as multi-functional drug delivery systems *Biomaterials* **34** 6068–81

[125] Gillies E R and Fréchet J M 2003 A new approach towards acid sensitive copolymer micelles for drug delivery *Chem. Commun.* **2003** 1640–1

[126] Guk K, Lim H, Kim B, Hong M, Khang G and Lee D 2013 Acid-cleavable ketal containing poly (β-amino ester) for enhanced siRNA delivery *Int. J. Pharmaceut.* **453** 541–50

[127] Shim M S and Kwon Y J 2012 Ketalized poly (amino ester) for stimuli-responsive and biocompatible gene delivery *Polym. Chem.* **3** 2570–7

[128] Yang Y Q *et al* 2013 pH-sensitive micelles self-assembled from multi-arm star triblock co-polymers poly (ε-caprolactone)-b-poly (2-(diethylamino) ethyl methacrylate)-b-poly (poly (ethylene glycol) methyl ether methacrylate) for controlled anticancer drug delivery *Acta Biomater.* **9** 7679–90

[129] Binauld S and Stenzel M H 2013 Acid-degradable polymers for drug delivery: a decade of innovation *Chem. Commun.* **49** 2082–102

[130] Zhou Z, Li L, Yang Y, Xu X and Huang Y 2014 Tumor targeting by pH-sensitive, biodegradable, cross-linked N-(2-hydroxypropyl) methacrylamide copolymer micelles *Biomaterials* **35** 6622–35

[131] Nakamura H *et al* 2014 Two step mechanisms of tumor selective delivery of N-(2-hydroxypropyl) methacrylamide copolymer conjugated with pirarubicin via an acid-cleavable linkage *J. Controlled Release* **174** 81–7

[132] Liu C, Liu F, Feng L, Li M, Zhang J and Zhang N 2013 The targeted co-delivery of DNA and doxorubicin to tumor cells via multifunctional PEI–PEG based nanoparticles *Biomaterials* **34** 2547–64

[133] Sun T-M *et al* 2014 Cancer stem cell therapy using doxorubicin conjugated to gold nanoparticles via hydrazone bonds *Biomaterials* **35** 836–45

[134] Dong D-W, Xiang B, Gao W, Yang Z-Z, Li J-Q and Qi X-R 2013 pH-responsive complexes using prefunctionalized polymers for synchronous delivery of doxorubicin and siRNA to cancer cells *Biomaterials* **34** 4849–59

[135] Etrych T, Šubr V, Laga R, Říhová B and Ulbrich K 2014 Polymer conjugates of doxorubicin bound through an amide and hydrazone bond: impact of the carrier structure onto synergistic action in the treatment of solid tumours *European Journal of Pharmaceutical Sciences* **58** 1–12

[136] Ding M *et al* 2013 Toward the next-generation nanomedicines: design of multifunctional multiblock polyurethanes for effective cancer treatment *ACS Nano* **7** 1918–28

[137] Zhou L *et al* 2012 The degradation and biocompatibility of pH-sensitive biodegradable polyurethanes for intracellular multifunctional antitumor drug delivery *Biomaterials* **33** 2734–45

[138] Heller J, Barr J, Ng S Y, Abdellauoi K S and Gurny 2002 Poly (ortho esters): synthesis, characterization, properties and uses *Adv. Drug Deliv. Rev.* **54** 1015–39

[139] Wang X *et al* 2012 A novel delivery system of doxorubicin with high load and pH-responsive release from the nanoparticles of poly (α, β-aspartic acid) derivative *European Journal of Pharmaceutical Sciences* **47** 256–64

[140] Koutroumanis K P, Holdich R G and Georgiadou S 2013 Synthesis and micellization of a pH-sensitive diblock copolymer for drug delivery *Int. J. Pharmaceut.* **455** 5–13

[141] Thambi T, Deepagan V, Yoo C K and Park J H 2011 Synthesis and physico-chemical characterization of amphiphilic block copolymers bearing acid-sensitive orthoester linkage as the drug carrier *Polymer* **52** 4753–9

[142] Tang R, Ji W, Panus D, Palumbo R N and Wang C 2011 Block copolymer micelles with acid-labile ortho ester side-chains: synthesis, characterization and enhanced drug delivery to human glioma cells *J. Controlled Release* **151** 18–27

[143] Ding C, Gu J, Qu X and Yang Z 2009 Preparation of multifunctional drug carrier for tumor-specific uptake and enhanced intracellular delivery through the conjugation of weak acid-labile linker *Bioconjugate Chem.* **20** 1163–70

[144] Vallet-Regí M, Balas F and Arcos D 2007 Mesoporous materials for drug delivery *Angew. Chem. Int. Ed.* **46** 7548–58

[145] Yang P, Gai S and Lin J 2012 Functionalized mesoporous silica materials for controlled drug delivery *Chem. Soc. Rev.* **41** 3679–98

[146] Yang K-N, Zhang C-Q, Wang W, Wang P C, Zhou J-P and Liang X-J 2014 pH-responsive mesoporous silica nanoparticles employed in controlled drug delivery systems for cancer treatment *Cancer Biol. Med.* **11** 34

[147] Wang Y *et al* 2014 Mesoporous silica nanoparticles in drug delivery and biomedical applications *Nanomedicine* **11** 313–27

[148] Slowing I I, Trewyn B G, Giri S and Lin V Y 2007 Mesoporous silica nanoparticles for drug delivery and biosensing applications *Adv. Funct. Mater.* **17** 1225–36

[149] Slowing I I, Vivero-Escoto J L, Wu C-W and Lin V S-Y 2008 Mesoporous silica nanoparticles as controlled release drug delivery and gene transfection carriers *Adv. Drug Deliv. Rev.* **60** 1278–88

[150] Yang Q *et al* 2005 pH-responsive carrier system based on carboxylic acid modified mesoporous silica and polyelectrolyte for drug delivery *Chem. Mater.* **17** 5999–6003

[151] Muhammad F *et al* 2011 pH-triggered controlled drug release from mesoporous silica nanoparticles via intracelluar dissolution of ZnO nanolids *J. Am. Chem. Soc.* **133** 8778–81

[152] Hong C-Y, Li X and Pan C-Y 2009 Fabrication of smart nanocontainers with a mesoporous core and a pH-responsive shell for controlled uptake and release *J. Mater. Chem.* **19** 5155–60

[153] Chen L, Zhang Z, Yao X, Chen X and Chen X 2015 Intracellular pH-operated mechanized mesoporous silica nanoparticles as potential drug carries *Micropor. Mesopor. Mat.* **201** 169–75

[154] Gan Q *et al* 2012 Endosomal pH-activatable magnetic nanoparticle-capped mesoporous silica for intracellular controlled release *J. Mater. Chem.* **22** 15960–8

[155] Lee J O, Oh K T, Kim D and Lee E S 2014 pH-sensitive short worm-like micelles targeting tumors based on the extracellular pH *J. Mater. Chem.* **B 2** 6363–70

[156] Raghupathi K, Li L, Ventura J, Jennings M and Thayumanavan S 2014 pH responsive soft nanoclusters with size and charge variation features *Polym. Chem.* **5** 1737–42

[157] Varkouhi A K, Scholte M, Storm G and Haisma H J 2011 Endosomal escape pathways for delivery of biologicals *J. Controlled Release* **151** 220–8

[158] Du J-Z, Du X-J, Mao C-Q and Wang J 2011 Tailor-made dual pH-sensitive polymer–doxorubicin nanoparticles for efficient anticancer drug delivery *J. Am. Chem. Soc.* **133** 17560–3

[159] Kelley E G, Albert J N, Sullivan M O and Epps T H III 2013 Stimuli-responsive copolymer solution and surface assemblies for biomedical applications *Chem. Soc. Rev.* **42** 7057–1

[160] Watkins K A and Chen R 2015 pH-responsive, lysine-based hydrogels for the oral delivery of a wide size range of molecules *Int. J. Pharmaceut.* **478** 496–503

[161] Ding X *et al* 2014 Hydrazone-bearing PMMA-functionalized magnetic nanocubes as pH-responsive drug carriers for remotely targeted cancer therapy *in vitro* and *in vivo* *ACS Appl. Mater. Interf.* **6** 7395–407

[162] Liu Y, Gao F-P, Zhang D, Fan Y-S, Chen X-G and Wang H 2014 Molecular structural transformation regulated dynamic disordering of supramolecular vesicles as pH-responsive drug release systems *J. Controlled Release* **173** 140–7

[163] Yu J, Deng H, Xie F, Chen W, Zhu B and Xu Q 2014 The potential of pH-responsive PEG-hyperbranched polyacylhydrazone micelles for cancer therapy *Biomaterials* **35** 3132–44

[164] Kang Y, Zhang X-M, Zhang S, Ding L and Li B-J 2015 pH-responsive dendritic polyrotaxane drug-polymer conjugates forming nanoparticles as efficient drug delivery system for cancer therapy *Polymer Chem.* **6** 2098–107

[165] Wang Y *et al* 2014 PEG–PCL based micelle hydrogels as oral docetaxel delivery systems for breast cancer therapy *Biomaterials* **35** 6972–85

[166] Tu F and Lee D 2014 Shape-changing and amphiphilicity-reversing Janus particles with pH-responsive surfactant properties *J. Am. Chem. Soc.* **136** 9999–10006

[167] Gui W, Wang W, Jiao X, Chen L, Wen Y and Zhang X 2014 Dual-cargo selectively controlled release based on a ph-responsive mesoporous silica system *ChemPhysChem* **16** 607–13

[168] He X, Li J, An S and Jiang C 2013 pH-sensitive drug-delivery systems for tumor targeting *Ther. Deliv.* **4** 1499–510

[169] Yu H *et al* 2014 Reversal of lung cancer multidrug resistance by pH-responsive micelle-plexes mediating co-delivery of siRNA and paclitaxel *Macromol. Biosci.* **14** 100–9

[170] Grainger S J and El-Sayed M E 2010 Stimuli-sensitive particles for drug delivery *Biol. Respons. Hybrid Biomater.* **2010** 171–90

[171] Thamphiwatana S, Fu V, Zhu J, Lu D, Gao W and Zhang L 2013 Nanoparticle-stabilized liposomes for pH-responsive gastric drug delivery *Langmuir* **29** 12228–33

[172] Gogoi N and Chowdhury D 2014 Novel carbon dot coated alginate beads with superior stability, swelling and pH responsive drug delivery *J. Mater. Chem.* B **2** 4089–99

[173] Rajpoot P, Bali V and Pathak K 2012 Anticancer efficacy, tissue distribution and blood pharmacokinetics of surface modified nanocarrier containing melphalan *Int. J. Pharmaceut.* **426** 219–30

[174] Torchilin V P 2014 Multifunctional, stimuli-sensitive nanoparticulate systems for drug delivery *Nat. Rev. Drug Discovery* **13** 813–27

[175] Koetting M C and Peppas N A 2014 pH-responsive poly (itaconic acid-co-N-vinyl-pyrrolidone) hydrogels with reduced ionic strength loading solutions offer improved oral delivery potential for high isoelectric point-exhibiting therapeutic proteins *Int. J. Pharmaceut.* **471** 83–91

[176] Choi S R *et al* 2014 Polymer-coated spherical mesoporous silica for pH-controlled delivery of insulin *J. Mater. Chem.* B **2** 616–9

[177] Nowag S and Haag R 2014 pH-responsive micro and nanocarrier systems *Angew. Chem. Int. Ed.* **53** 49–51

[178] Singh I *et al* 2013 Low levels of copper disrupt brain amyloid-β homeostasis by altering its production and clearance *Proc. Natl Acad. Sci.* **110** 14771–6

[179] Treiber C *et al* 2009 Cellular copper import by nanocarrier systems, intracellular availability and effects on amyloid β peptide secretion *Biochemistry* **48** 4273–84

[180] Ejima H *et al* 2013 One-step assembly of coordination complexes for versatile film and particle engineering *Science* **341** 6142

[181] Shi J *et al* 2014 A tumoral acidic pH-responsive drug delivery system based on a novel photosensitizer (fullerene) for *in vitro* and *in vivo* chemo-photodynamic therapy *Acta Biomater.* **10** 1280–91

Chapter 3

Enzyme-responsive nanocarriers

3.1 Introduction

Enzymes are a major component of the bio-nanotechnology toolbox. The concept of enzymes was conceived more than 100 years ago by Paul Ehrlich. Enzymes have attracted the attention of a large number of researchers thanks to their unique properties, including exceptional biorecognition capability, and particular catalytic functions that lead to very selective action on their specific substrates [1]. Enzyme-catalyzed reactions are highly efficient and selective for specific substrates under relatively mild conditions [1]. When the enzymatic activity is associated with a particular tissue or the enzyme is found at higher concentrations at the target site, an enzyme-responsive nanomaterial can be programmed to deliver and release drugs via enzymatic conversion of the carrier. Moreover, detection of enzyme activity can be an extremely useful tool in diagnostics, since dysregulation of enzyme expression is a characteristic feature of numerous diseases. The design of a DDS to take advantage of the specific enzyme activity can be based on either a physical or chemical mode of action. In the chemical mode of action, it is possible to program the nanomaterials to release their drug cargo using the degradation of the polymeric shell that happens when the nanomaterial encounters the enzyme. In this method, transformation of the carrier by the enzyme can also release the therapeutic molecules, therefore paving the way to design multimodal nanomedicines with synergistic effects. This strategy can also be applied to self-assembled nanomaterials that encapsulate cargo, for example amphiphile-based NP [2]. Based on this strategy, degradation or transformation of the carrier by the enzyme can release the therapeutic molecules. In the physical mode, the enzyme-responsive NP can be designed so that their macro-scale structure is altered by the enzyme, which releases the cargo. According to this method, the nanomaterial itself is not responsive to biocatalysis, but its surface can be modified with molecules that generate a change in

doi:10.1088/978-1-6817-4257-1ch3   3-1

the physical properties of the nanoparticle solution upon enzymatic transformation. All in all, the main advantages of enzyme-responsive materials (ERM) can be summarized as follows:
- enzymes are natural products;
- enzyme reactions are highly specific;
- a large number of enzyme/substrate pairs are available in the biomedical literature;
- polymer materials can be designed to respond to specific enzymes with reversible or irreversible changes to their chemical structure;
- the chemistry for incorporating enzyme-sensitive functional groups is available;
- dysregulated enzymes are often present in diseased tissue;
- typical enzyme-sensitive biomolecules (DNA and peptides) are very versatile and it is easy to tune the sequence to match each enzyme specificity;
- a single material can be responsive to more than one enzyme.

However, there are also several limitations:
- enzymes generally work in specific physiological conditions;
- with the exception of biomolecular polymers, polymers are not generally enzyme-sensitive and require separate attached functionalities;
- polymers should be stable in complex biological environments and resistant to non-specific enzyme attack.

3.2 Immobilized biocatalysts

Covalent coupling of an enzyme to a smart polymer, in solution or in a hydrogel, can significantly change its activity and can alter substrate access to the active site due to changes in the polymer and protein conformations. The transition between the soluble and insoluble states of smart polymers has been exploited in the development of reversibly soluble biocatalysts. These biocatalysts can accelerate an enzymatic reaction in their soluble state and hence they can be utilized in reactions with insoluble substrates. As soon as the reaction is complete and the products are released from the protein, the conditions are changed to allow the catalyst to be utilized in the next cycle after redissolution. A wide range of smart polymers with pH-controlled solubility have been used for the increase of reversibly soluble biocatalysts [3–12].

3.3 Enzyme-responsive materials in drug delivery

The products of the reaction catalyzed by an immobilized enzyme contained in smart hydrogels can themselves cause the gel phase transition. It is then possible to transform a chemical signal such as the presence of the substrate into an environmental signal, such as a pH change [6], or a mechanical signal, such as shrinking or swelling of the smart gel. An enzyme-responsive material (ERM) can be designed as a system that undergoes macroscopic changes of its physical/chemical properties upon the catalytic action of an enzyme. The response mechanism of ERM requires an enzyme-sensitive part, and also a second component that is used to allow changes between the interactions inside the material that can lead to macroscopic transitions. Common ERM are reviewed below.

3.4 Common enzyme-responsive materials

In this section, the major systems that can be used as DDS are classified according to the function that is altered by enzyme action. Based on the chemical and physical strategies of the delivery system, the ERM can be categorized into two different groups: hydrogels and NP (figure 3.1). Enzymic reorganization and the cleavage of functionalized ligands are their main properties. In other words, the specified enzymes change the physiochemical properties of the nanomaterial [13].

3.4.1 Polymeric materials

Hydrogels

The main polymers used in the preparation of enzyme-responsive hydrogels can be classified as natural polymers (such as polypeptides and polysaccharides) and

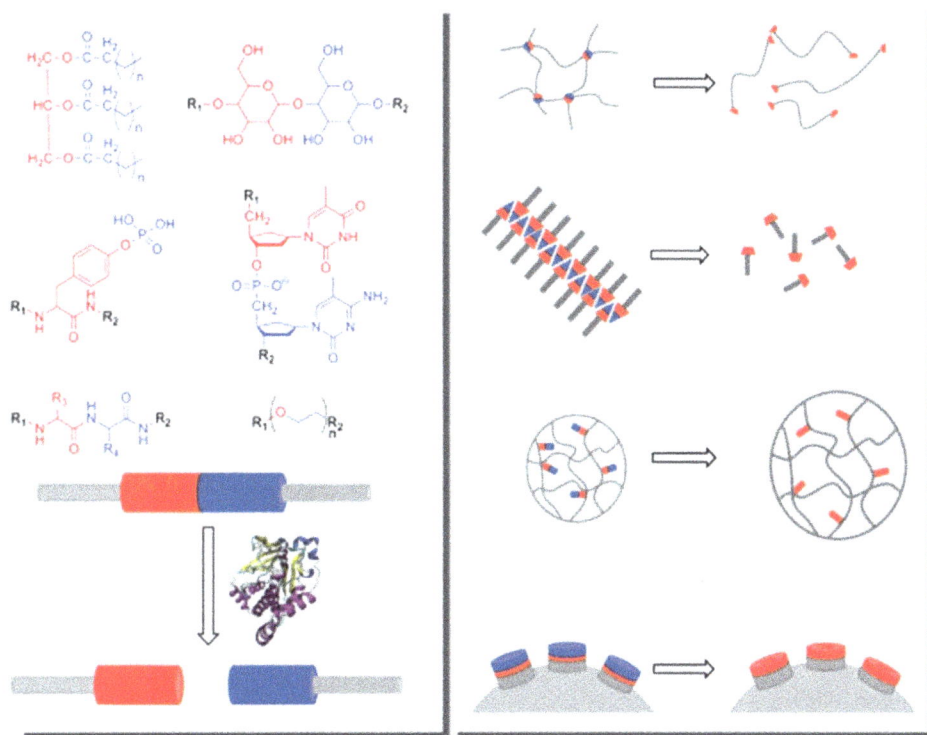

Figure 3.1. Left: chemical structure of common enzymes and examples of molecules used as enzyme substrates. Right: schematic description of enzyme-responsive systems; from top to bottom: degradable hydrogels, supramolecular hydrogels, triggered swelling hydrogels and a silica nanocontainer. Reproduced with permission from [13]. Copyright 2014 by The Royal Society of Chemistry.

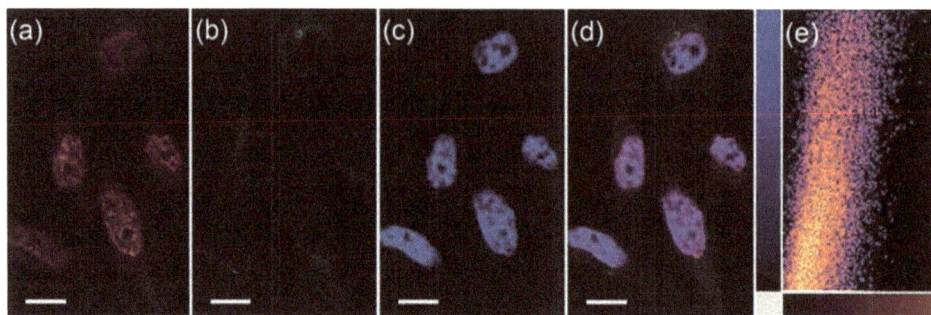

Figure 3.2. The confocal analysis results for HeLa cells incubated with Hep(DOX)SN gel for 20 h. (a) Delivered drug, (b) lysosomes, (c) nucleus, (d) overlay and (e) the colocalization analysis results for the nucleus/DOX. DAPI (blue), DOX (red) and LysoTracker green DND-26 (green). The scale bars show 10 μm. Reproduced with permission from [24]. Copyright 2012 by The Royal Society of Chemistry.

synthetic polymers (such as poly(N-isopropylacrylamide) (PNIPAAm) and PEG/poly(ethylene oxide) (PEO)) [14, 15]. The natural polymers are degraded directly. Synthetic polymers, by contrast, require the attachment of enzyme-sensitive moieties such as short peptide sequences. Polymer hydrogels can be prepared so that they are susceptible to degradation by enzymes such as elastase and metalloproteinases [16–19]. Enzymes have been used to control the morphology of hydrogel particles as well as destroy them. Using enzymes such as trypsin, thermolysin and elastase, the degree of swelling in polymers can be altered, with drugs being released accordingly [20–23].

In one study, Su *et al* [24] synthesized a new enzyme-responsive hydrogel using an N-hydroxyimide–heparin conjugate and glucose oxidase (GOx). In this system, the GOx-mediated radical polymerization created flexible hydrogels with excellent physical and chemical characteristics. Confocal microscopy depicted the intercellular distribution of the released cargo and demonstrated that it was possible to deliver the drug from the developed hydrogel thanks to the heparin-specific enzymatic cleavage reaction. According to figures 3.2(a)–(d), a diffuse signal for the released drug occurred in the nucleus of the targeted cells after the incubation of HeLa cells with the Hep(DOX)SN gel for 20 h. The colocalization analysis results for the nucleus/DOX are presented in figure 3.2(e), where it is seen that a considerable proportion of DOX translocated to the nucleus by escaping from the cytoplasm. However, in HepG2 cells the greater proportion of released cargo remained trapped in the cytoplasm. Su *et al* concluded that the Hep(DOX)SN gel may release the drug in a controlled fashion by responding to the environmental levels of heparanase.

Nanoparticles
Polymer hydrogels have greater than 99% water content and are characterized by a crosslinked network within the polymer. The main polymers used in enzyme-responsive hydrogels can be classified as naturally occurring (e.g. polypeptides and polysaccharides) and synthetic (e.g. poly N-isopropylacrylamide (PNIPAAm) and PEG/PEO) [14, 15]. Natural biopolymers are degraded directly. Synthetic polymers,

by contrast, require the attachment of enzyme-sensitive moieties such as short peptide sequences. In this fashion, polymer hydrogels can be degraded by enzymes such as elastase and metalloproteinases [16–19]. Not only can enzymes destroy the polymer hydrogel structure, but they can also control the morphology of the hydrogel particles. Using enzymes such as trypsin, thermolysin and elastase, the degree of swelling in polymers can be altered, leading to drug release [20–23]. Klinger *et al* designed an enzyme-degradable hydrogel by crosslinking poly (PAAm) with dextran methacrylate (Dex-MA), and this was partially biodegradable by enzymatic cleavage of the methacryl-functionalized polysaccharide chains [25].

Enzyme-responsive NP have also been studied for triggered release of cargo. 1,4-Hydroxymandelic acid was used as a framework for attaching an enzyme-cleavable group such as a phenylacetic acid residue, which was used as a substrate for cleavage by bacterial penicillin G amidase. A twin-arginine translocation (Tat) peptide was the peptide carrier and acted as a protein-transduction domain, while nalidixic acid was the antibacterial cargo. The enzyme-activated system exhibited a minimum inhibitory concentration (MIC) value against *E. coli* that was about 70 times lower (1.9 μM) than that for free nalidixic acid (138 μM) [26–29].

Gu and Tang devised a protein nanocapsule interwoven with an enzyme-degradable polymeric network [30–33]. Proteins were functionally encapsulated using a cocoon-like polymeric nanocapsule formed by interfacial polymerization. The nanocapsule was crosslinked by peptides that could be proteolyzed by proteases, thus releasing the protein cargo. The protease-mediated degradation process could be controlled in a spatiotemporal fashion through modification of the peptide crosslinker with photolabile moieties. They demonstrated the utility of this approach through the cytoplasmic delivery of the apoptosis inducing caspase-3 to cancer cells. Organic and inorganic hybrid NP can be composed of organic polymers or inorganic colloids. For instance, in lipid–polymer hybrid NP, biodegradability, mechanical firmness, increased surface area-to-volume ratio and narrow size distribution can be provided by solid cores [34, 35]. The nanomaterials used in enzyme-responsive DDS are often responsive to oxidoreductases or hydrolases [6, 36, 37].

3.4.2 Oxidoreductase-responsive nanomaterials

Oxidoreductases have been identified as therapeutic targets thanks to their central role in oxidative stress and, moreover, monoamine oxidases are responsible for the activation of neurotransmitters [38]. Based on the general definition, oxidoreductase refers to a type of enzyme responsible for catalytic reactions involving oxidation/reduction in living systems. In such systems, the transfer of electrons between biological molecules occurs via an enzyme cofactor that can function as an electron carrier (e.g. NAD^+ or $NADP^+$). In this reaction, the electron acceptor substrate is an oxidant, while the electron donor substrate is called a reductant [39]. Furthermore, oxidoreductase enzymes can be categorized into several classes, namely oxidases, peroxidases, oxygenases, reductases and dehydrogenases [40]. Most recent research has been performed on oxidoreductase enzymes with applications ranging from

DDS through disease diagnostics to bio-imaging probes, and so on. In the following sections, the major groups of oxidoreductase enzymes incorporated in DDS are discussed in detail.

Glucose-responsive nanomaterials

It is generally accepted that glucose content plays a significant role in serious diseases such as diabetes mellitus, where an imbalance in glucose levels causes kidney failure, blindness, heart attacks and death. With this in mind, the detection and control of glucose levels in the blood is required for both diagnosis and treatment of diabetes mellitus [41]. In recent years, DDS employing an implantable polymer matrix that can function as self-regulated insulin delivery systems through molecularly programmed release have attracted much attention [42–46]. In these DDS, hydrogels that are sensitive to glucose improve glucose metabolism by releasing insulin in a controlled manner. Specific enzymes may change the local pH level in the hydrogel environment and the insulin is released accordingly [42–46]. The diffusion of increased glucose into the hydrogel and the conversion of glucose into gluconic acid is the main mechanism in these kinds of DDS. This reaction is usually accompanied by the pH changes at the specific site; as a result, the polymeric hydrogel swells, the diffusion of insulin within the polymeric matrix is triggered, and release from the hydrogel is improved. In the treatment of diabetes, hydrogels that respond to glucose by swelling have the potential to be self-regulating insulin release systems [47–51].

A study modeled the release of insulin in a model of streptozotocin (STZ)-induced diabetic mice. It showed that the insulin-loaded nanocarriers can act as a diabetes treatment [38, 52]. In another important study, Napoli *et al* [53] synthesized an encapsulated glucose oxidase in polymeric vesicles. They used polypropylene sulfide (PSS) and polyethylene glycol (PEG) precursors as building blocks in nanocarriers [54]. The results showed that the presence of a moderate level of glucose destabilized the polymeric shell and allowed interaction with glucose oxidase. The developed DDS can be used for the transformation of glucose to gluconolactone, with hydrogen peroxide being produced as a by-product. The hydrogen peroxide is an oxidizing agent and can produce sulfoxides and sulfones by oxidizing thioethers; as a result, the nanocarriers are destabilized.

3.4.3 Hydrolase-responsive nanomaterials

Hydrolases are a group of enzymes that are widely applied as effector biomolecules in enzyme-activated DDS [2]. These comprise different types of enzymes, e.g. proteases, which often exist inside cells and will act on nanocarriers after endocytosis [55]. Trypsin can degrade protamines and chondroitin sulfate to release the encapsulated cargo [56]. Phospholipase activity is increased in infected areas and in different types of cancer, such as pancreatic and prostate [56]. Elastase in particular increases during inflammation and can digest the collagen and elastin components of NP to cause drug release [57]. The carrier nanomaterial can be digested by the hydrolase whenever the concentration of both species is high enough. This strategy can reduce the side-effects of drugs and, moreover, the products of the enzymatic digestion can be designed to be a therapeutic molecule, for example by destabilizing the cell shell to facilitate the penetration of the drug [2]. The following section reviews the different kinds of hydrolase-responsive nanocarriers.

Protease-responsive nanomaterials

Proteases are ideal enzymes for designing smart DDS, because of their highly controlled reactivity and their involvement in biological processes, such as DNA replication, transcription, wound repair, blood coagulation and stem cell mobilization [58]. Proteases are produced by both normal cells and tumor cells and they can cleave proteins, turning them into smaller peptides or amino acids. In addition, multimeric and multicatalytic proteases associate into intracellular protein complexes called proteasomes, which are essential for biological processes [59]. A polymeric carrier containing the cargo can be connected to the protease-sensitive peptide. Consequently, when the first type of protease-sensitive NP accumulate in a tumor site due to the EPR effect, the second type of NP containing the effector protease can be administered, leading to localized drug release with minimum nonspecific toxicity. Proteases are also over-expressed in some diseases and can be used to release drugs from nanocarriers at specific locations within the cells [60]. By so doing, the toxic effects of drugs are reduced and the therapeutic effects are enhanced. It should be said that this effect is mainly achieved because the therapeutic effects of the drugs only apply after the accumulation of the NP inside the cell via endocytosis. For example, the results of a study by Vicent *et al* [61] showed that a specific peptide sequence can control the release of aminoglutethimide (AGM) and DOX efficiently when used as a linker to N-(2-hydroxypropyl) methacrylamide (HPMA) NP.

A peptide structure that is activated by a protease and might have an application in targeting against HIV is shown in figure 3.3. Satchi *et al* [62] reported a synergistic effect from the chemotherapy drug DOX and a form of endocrine therapy, aminoglutethimide (AGM), which showed a markedly enhanced cytotoxicity *in vitro* when both compounds were delivered via the same polymer [63].

The expression of matrix metalloproteinases (MMP) is often specific to the tumor microenvironment and/or the inside of cancer cells [64, 65]. The substrate specificity

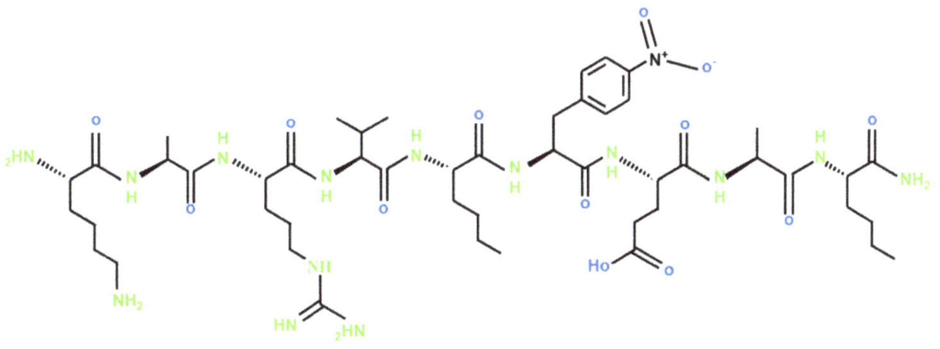

Figure 3.3. Chemical structure of a peptide substrate (HIV protease substrate IV) that can be cleaved by a specific protease.

of MMP involves the degradation of specific peptide sequences, which control aspects of cellular behavior. Jiang *et al* developed a peptide-modified nanoparticle DDS triggered by MMP-2 and MMP-9 that can be used to deliver Cy5 fluorophore or gadolinium, or both species at once [66, 67].

de la Rica *et al* showed that proteases can reduce the toxic side-effects of released drugs in addition to increasing their therapeutic effects [2]. For example, an innovative nanoparticle-in-microparticle delivery system (NiMDS) was developed by Imperiale *et al* [68]. This system comprised a pure drug nanocrystal composed of the protease inhibitor, indinavir free base, which was produced by nanoprecipitation and was encapsulated within mucoadhesive polymeric microparticles. This DDS could be a crucial step towards overcoming the current problem of patient non-compliance with protease inhibitor therapy, as the use of this platform could reduce the dose and the frequency of administration of protease inhibitors.

Phospholipase-responsive nanomaterials
Phospholipases have been considered to be a potential target enzyme in therapeutics owing to their unique role in inflammatory diseases and infections. These enzymes include phospholipase A2 (PLA2), which is up-regulated in inflammatory and infectious diseases. It has also been suggested that PLA2 plays a role in the host defense mechanism against tumors because of its abnormally high concentration in the invading zone of tumors [69, 70].

The self-assembly of phospholipids in water can form liposomes. The lipid composition specifies their chemical and physical properties, such as permeability and surface charge, allowing the fabrication of nanocarriers with tailored release properties [71]. Liposomes are biocompatible and easy to fabricate, and so have been widely utilized in DDS [72]. These organic carriers can encapsulate the molecule of interest and can also work as a pro-drug when cytotoxic lipids are formed via enzymatic digestion of the carrier (antitumor ether lipids or AEL). This cytotoxicity could be used as an anticancer drug, for example, minimizing the hemolytic side-effect of AEL [73].

Trypsin-responsive nanomaterials
Trypsin, one of the most important digestive proteinases, is involved in the process of digestion, which begins in the stomach and follows in the small intestine where a slightly alkaline environment elevates trypsin's enzymatic activity [74]. In the digestive process, trypsin helps the other proteinases to break down dietary protein molecules into their constituent amino acids and peptides [75]. Radhakrishnan *et al* [56] demonstrated the ability of either trypsin or hyaluronidase enzymes, which are over-expressed simultaneously under pathological conditions, to disintegrate capsules fabricated from protamine and chondroitin sulphate. When these nanocapsules were exposed to pH 7.4, the crosslinking was maintained and the drug molecules were rapidly released in the presence of either one of the triggering enzymes [76]. Wang *et al* synthesized a four-component nanocomposite formed from trypsin-immobilized polyaniline-coated Fe_3O_4/carbon nanotubes for highly efficient protein digestion. This biocomposite offered considerable promise for protein

analysis due to its high magnetic responsivity and excellent dispersibility. Because the enzyme-immobilized nanocomposite can be prepared using a simple, low-cost two-step deposition approach, it may find a wide range of biological applications, including proteome research [77].

Glycosidase-responsive nanomaterials
Glycosidase enzymes catalyze the hydrolysis of carbohydrates to yield small sugar molecules and can be utilized for developing smart DDS. Glycosidase, a carbohydrate-hydrolysing enzyme, is found in such important sites as cancer cells, HIV, metastases, inflammation and infections. In this regard, studies have demonstrated that the hydrolysis of carbohydrates increases significantly (85%) in common carcinomas, e.g. lung, breast, pancreas, ovarian, stomach, uterine, osteosarcoma and multiple myeloma. As a result, sugar-based NP can be applied as a drug carrier in an appropriate manner [2]. For instance, Bernados *et al* [78] synthesized a silica mesoporous composite containing a glycosidase responsive moiety, a MCM-41-based material and some derivatives of saccharides (called Glucidex 29, 39, 47) for a gate-like hybrid release system. They also reported that delivery of DOX to Hela and LLC-PK1 cells increases drastically in the presence of β-D-galactosidase [78]. Although a large number of research studies have focused on these kinds of nanomaterials, it seems that more studies need to be performed to enhance the efficiency and selectivity of glycosidase-responsive materials in different DDS.

Elastase-responsive nanomaterials
Another type of protease enzyme (peptidase) is elastase, which degrades and digests proteins such as elastin, an elastic protein that features in the lungs and some other organs and governs the mechanical properties of connective tissue. Elastases may belong to different classes: serine proteinases, cysteine proteinases and metallo-proteinases. Mammalian elastases mostly occur in the pancreas and in phagocytic cells. In the non-mammalian elastase group, there is a large diversity of bacterial metallo- and serine elastases. There are many natural (proteins) and synthetic elastase inhibitors. Elastases play a pathologic role in pulmonary emphysema, cystic fibrosis, infections, inflammation and atherosclerosis [79]. Predicting the elastinolytic behavior of a proteinase is difficult, mainly because these enzymes have variable catalytic and substrate binding sites. These substrates are also hydrolyzed by non-elastinolytic enzymes, or they may be resistant to elastases [80].

Kim *et al* [81] claimed that the neutrophil elastase enzyme might play an important role in the progression of ventilator-induced lung injury (VILI). They showed that the mechanisms underlying the effect of neutrophil elastase on lung injury involve increased neutrophil elastase activity, resulting in an increase to pulmonary vascular permeability. Their results indicated that pretreatment with sivelestat (an elastase inhibitor) can prevent the progression of lung injury caused by neutrophil elastase, although further clinical studies are needed to confirm the results reported. Sun *et al* [82] found that serum levels of neutrophil elastase are increased in patients with influenza-induced encephalitis (IE) compared to those with uncomplicated influenza, which suggests that cerebral endothelial damage in the development of IE is mediated by neutrophil elastase. This study implies that anti-elastase agents may be an effective therapy for IE, but this claim needs further confirmation.

All in all, it seems that using different kinds of elastases to trigger cargo release may potentially have a role in DDS, especially in respect of lung diseases. For example, some of the most important drugs used to treat emphysema are peptidyl carbamates, α1-antitrypsin and Eglin C. These drugs prevent the remodeling of lung tissue by reducing free-elastase activity in the lungs.

3.4.4 Other enzyme-responsive nanomaterials

Recently, bacterial enzymes (BE) have attracted the attention of researchers who have proposed that these kinds of enzymes can be used to release antibiotics inside cells. Xiong *et al* [83] focused on this new research field and treated murine salmonellosis using gentamicin in a silica xerogel-based DDS. In their system for targeted DDS, the polyphosphoester core of the antibiotic-loaded carrier is degraded by BE.

In another study, Rao *et al* [84, 85] developed a DDS using an azoreductase-responsive nanomaterial to trigger the release of drugs in colonic disease. This enzyme is produced by microbial flora and is present in the human colon. The researchers assembled azobenzene and an amphiphilic diblock copolymer by forming a covalent coupling between them. In the presence of the azoreductase and nicotinamide adenine dinucleotidephosphate (NADPH), the vehicle disinte-grated, releasing its drug.

Kallikreins are serine-based enzymes associated with poor clinical prognoses for human cancer. Human kallikrein 3 (hK3), a prostate-specific antigen, is the most popular one for use in targeting prostate cancer. Sex-steroid hormones regulate the gene expression of kallikrein in prostate tissue [86]. However, post-translational modification controls the chemical actions of this irreversible protease. Nonetheless, recent investigations have indicated that hK can both advance and prevent tumor progression, with this heavily dependent on hormone levels and tissue type [87]. López-Otín *et al* [88] investigated the tumor-suppressing role of hK3, hK8, hK9, hK10, hK13 and hK 14, concluding that the overexpressed hK could be used as feasible targets in DDS.

The enzyme urease has been employed to trigger the disassembly of gold NP through the hydrolysis of urea to yield ammonium and bicarbonate ions. This DDS was utilized in the design of a highly sensitive enzyme-linked immunosorbent assay (ELISA) [89].

Protease-sensitive materials have high potential for use in DDS, so it is hoped that future studies will discover further aspects of them, thanks to progress in such research fields as molecular modeling and protein engineering. For instance, researchers could focus on using protease-responsive nanomaterials to target the mitochondria in patients' hematological cells, and oxidoreductase-responsive DDS to target chloroplasts and the nucleus. The latest achievements of researchers using several classes of enzyme-responsive micro/nano materials are summarized in table 3.1.

Table 3.1. Examples of enzyme-responsive nanomaterials and their applications.

Enzyme main group	Enzyme subgroup	Micro- and nanomaterial	Linked disease	Application
Hydrolases	α-Chymotrypsin	Porous CaCO$_3$ microparticles	—	DNA/PLL capsules with high loading capacity for dextran (high transport efficiency of drug) [90]
	Trypsin	Silver NP (AgNP)	—	Increasing the hydrolyzing capacity of bovine serum albumin [91]; quartz crystal microbalance (QCM) [92]
	Trypsin/gelatinase	Gold NP (AuNP)	—	Pancreatitis diagnosis and therapies [75]
	Proteases	Gold NP (AuNP)	Tumors, cancers	Optical biosensing platform for proteinase activity [93]
	Proteases (CAP); elastase	Liposomes	Prostate cancer	Drug delivery vehicles [60]
	Lipases (PLA$_2$)	Gold NP	—	Care diagnostics; high-throughput pharmaceutical screening
Oxidoreductase	Glucose oxidase	Chitosan nanogel-based capsules	Type 1 diabetes	Insulin-delivery systems [94]
	Glucose oxidase	Liposomes	—	Nanocontainers in sensing devices; DDS [53]
	Glucose	Double-layered nanogel	Diabetes	Insulin-delivery systems [95]
	Glucose oxidase	Gold NP (AuNP)	—	Glucose biosensor
	Peroxidase	Gold NP (AuNP)	—	ELISA [96]

References

[1] Hu J, Zhang G and Liu S 2012 Enzyme-responsive polymeric assemblies, nanoparticles and hydrogels *Chem. Soc. Rev.* **41** 5933–49

[2] de la Rica R, Aili D and Stevens M M 2012 Enzyme-responsive nanoparticles for drug release and diagnostics *Adv. Drug Deliv. Rev.* **64** 967–78

[3] Cong L, Kaul R, Dissing U and Mattiasson B 1995 A model study on Eudragit and polyethyleneimine as soluble carriers of α-amylase for repeated hydrolysis of starch *J. Biotechnol.* **42** 75–84

[4] Chen J P and Chang K C 1994 Immobilization of chitinase on a reversibly soluble–insoluble polymer for chitin hydrolysis *J. Chem. Technol. Biotechnol.* **60** 133–40

[5] Hoshino K *et al* 1997 Continuous simultaneous saccharification and fermentation of delignified rice straw by a combination of two reversibly soluble-autoprecipitating enzymes and pentose-fermenting yeast cells *J. Chem. Eng. Jpn* **30** 30–7

[6] Hoshino K *et al* 1996 Repeated utilization of β-glucosidase immobilized on a reversibly soluble–insoluble polymer for hydrolysis of phloridzin as a model reaction producing a water-insoluble product *J. Ferment. Bioeng.* **82** 253–8

[7] Sun Y, Jin X-H, Dong X-Y, Yu K and Zhou X 1996 Immobilized chymotrypsin on reversibly precipitable polymerized liposome *Appl. Biochem. Biotechnol.* **56** 331–9

[8] Chen J-P and Hsu M-S 1997 Preparations and properties of temperature-sensitive poly (N-isopropylacrylamide)–chymotrypsin conjugates *J. Mol. Catal. B: Enzym.* **2** 233–41

[9] Chilkoti A, Chen G, Stayton P S and Hoffman A S 1994 Site-specific conjugation of a temperature-sensitive polymer to a genetically engineered protein *Bioconjugate Chem.* **5** 504–7

[10] Chen G and Hoffman A S 1995 Graft copolymers that exhibit temperature-induced phase transitions over a wide range of pH *Nature* **373** 6509

[11] Shiroya T, Yasui M, Fujimoto K and Kawaguchi H 1995 Control of enzymatic activity using thermosensitive polymers *Colloid. Surf. B* **4** 275–85

[12] Nagayama H, Maeda Y, Shimasaki C and Kitano H 1995 Catalytic properties of enzymes modified with temperature-responsive polymer chains *Macromol. Chem. Phys.* **196** 611–20

[13] Hu Q, Katti P S and Gu Z 2014 Enzyme-responsive nanomaterials for controlled drug delivery *Nanoscale* **6** 12273–86

[14] Sperinde J J and Griffith L G 1997 Synthesis and characterization of enzymatically cross-linked poly (ethylene glycol) hydrogels *Macromolecules* **30** 5255–64

[15] Wichterle O and Lim D 1960 Hydrophilic gels for biological use *Nature* **185** 117–8

[16] van Dijk M, van Nostrum C F, Hennink W E, Rijkers D T and Liskamp R M 2010 Synthesis and characterization of enzymatically biodegradable PEG and peptide-based hydrogels prepared by click chemistry *Biomacromolecules* **11** 1608–14

[17] Yang J, Jacobsen M T, Pan H and Kopeček J 2010 Synthesis and characterization of enzymatically degradable PEG-based peptide-containing hydrogels *Macromol. Biosci.* **10** 445–54

[18] Lutolf M and Hubbell J 2003 Synthesis and physicochemical characterization of end-linked poly (ethylene glycol)-co-peptide hydrogels formed by Michael-type addition *Biomacromolecules* **4** 713–22

[19] Kim S and Healy K E 2003 Synthesis and characterization of injectable poly (N-isopropylacrylamide-co-acrylic acid) hydrogels with proteolytically degradable cross-links *Biomacromolecules* **4** 1214–23

[20] Aguilar M R and Román J 2014 *Smart Polymers and Their Applications* (Amsterdam: Elsevier)

[21] Thornton P D, Mart R J, Webb S J and Ulijn R V 2008 Enzyme-responsive hydrogel particles for the controlled release of proteins: designing peptide actuators to match payload *Soft Matter* **4** 821–7

[22] Patrick A G and Ulijn R V 2010 Hydrogels for the detection and management of protease levels *Macromol. Biosci.* **10** 1184–93

[23] McDonald T O, Qu H, Saunders B R and Ulijn R V 2009 Branched peptide actuators for enzyme responsive hydrogel particles *Soft Matter* **5** 1728–34

[24] Su T *et al* 2014 Glucose oxidase triggers gelation of N-hydroxyimide–heparin conjugates to form enzyme-responsive hydrogels for cell-specific drug delivery *Chemical Science* **5** 4204–9

[25] Klinger D, Aschenbrenner E M, Weiss C K and Landfester K 2012 Enzymatically degradable nanogels by inverse miniemulsion copolymerization of acrylamide with dextran methacrylates as crosslinkers *Polymer Chem.* **3** 204–16

[26] Lee M R, Baek K H, Jin H J, Jung Y G and Shin I 2004 Targeted enzyme-responsive drug carriers: studies on the delivery of a combination of drugs *Angew. Chem. Int. Ed.* **43** 1675–8

[27] Gu G *et al* 2013 PEG–co–PCL nanoparticles modified with MMP-2/9 activatable low molecular weight protamine for enhanced targeted glioblastoma therapy *Biomaterials* **34** 196–208

[28] Petros R A and DeSimone J M 2010 Strategies in the design of nanoparticles for therapeutic applications *Nat. Rev. Drug Discovery* **9** 615–27

[29] Davis M E 2008 Nanoparticle therapeutics: an emerging treatment modality for cancer *Nat. Rev. Drug Discovery* **7** 771–82

[30] Gu Z and Tang Y 2010 Enzyme-assisted photolithography for spatial functionalization of hydrogels *Lab Chip* **10** 1946–51

[31] Gu Z, Biswas A, Joo K-I, Hu B, Wang P and Tang Y 2010 Probing protease activity by single-fluorescent-protein nanocapsules *Chem. Commun.* **46** 6467–9

[32] Gu Z *et al* 2009 Protein nanocapsule weaved with enzymatically degradable polymeric network *Nano Lett.* **9** 4533–8

[33] Chien M-P *et al* 2013 Enzyme-directed assembly of nanoparticles in tumors monitored by *in vivo* whole animal imaging and *ex vivo* super-resolution fluorescence imaging *J. Am. Chem. Soc.* **135** 18710–3

[34] Zuo A *et al* 2008 Improved transfection efficiency of CS/DNA complex by co-transfected chitosanase gene *Int. J. Pharm.* **352** 302–8

[35] Schärtl W 2010 Current directions in core–shell nanoparticle design *Nanoscale* **2** 829–43

[36] Cooper S, Bamford C and Tsuruta T 1995 *Polymer Biomaterials in Solution, as Interfaces and as Solids: A Festrschrift Honoring the 60th Birthday of Dr Allan S Hoffman* (Utrecht: VSP)

[37] Matsuda T *et al* 1995 Association and activation of Btk and Tec tyrosine kinases by gp130, a signal transducer of the interleukin-6 family of cytokines *Blood* **85** 627–33

[38] Kundu J K and Surh Y-J 2010 Nrf2-Keap1 signaling as a potential target for chemo-prevention of inflammation-associated carcinogenesis *Pharmaceut. Res.* **27** 999–1013

[39] Eslahi H, Ghaffari-Moghaddam M, Khajeh M, Omay D, Zakipourrahimabadi E and Motalleb G 2014 General biography, structure and classification of enzymes *Res. Rev. Mater. Sci. Chem.* **3** 1–83

[40] Rix U, Fischer C, Remsing L L and Rohr J 2002 Modification of post-PKS tailoring steps through combinatorial biosynthesis *Nat. Prod. Rep.* **19** 542–80

[41] Wang J 2008 Electrochemical glucose biosensors *Chemical Reviews* **108** 814–25

[42] Jin R, Hiemstra C, Zhong Z and Feijen J 2007 Enzyme-mediated fast *in situ* formation of hydrogels from dextran–tyramine conjugates *Biomaterials* **28** 2791–800

[43] Brown L R, Edelman E R, Fischel-Ghodsian F and Langer R 1996 Characterization of glucose-mediated insulin release from implantable polymers *Journal of Pharmaceutical Sciences* **85** 1341–5

[44] Chandy T and Sharma C P 1992 Glucose-responsive insulin release from poly (vinyl alcohol)-blended polyacrylamide membranes containing glucose oxidase *J. Appl. Polym. Sci.* **46** 1159–67

[45] Kikuchi A and Okano T 2002 Pulsatile drug release control using hydrogels *Adv. Drug Deliv. Rev.* **54** 53–77

[46] Wu W and Zhou S 2013 Responsive materials for self-regulated insulin delivery *Macromol. Biosci.* **13** 1464–77

[47] Imanishi Y and Ito Y 1995 Glucose-sensitive insulin-releasing molecular systems *Pure Appl. Chem.* **67** 2015–22

[48] Kai G and Min Y 2001 AAc photografted porous polycabonate films and its controlled release system *J. Controlled Release* **71** 221–5

[49] Tang M, Zhang R, Bowyer A, Eisenthal R and Hubble J 2003 A reversible hydrogel membrane for controlling the delivery of macromolecules *Biotechnol. Bioeng.* **82** 47–53

[50] Podual K, Doyle F J III and Peppas N A 2000 Glucose-sensitivity of glucose oxidase-containing cationic copolymer hydrogels having poly (ethylene glycol) grafts *J. Controlled Release* **67** 9–17

[51] Miyata T, Uragami T and Nakamae K 2002 Biomolecule-sensitive hydrogels *Adv. Drug Deliv. Rev.* **54** 79–98

[52] Butterfield D A, Hardas S S and Lange M L B 2010 Oxidatively modified glyceraldehyde-3-phosphate dehydrogenase (GAPDH) and Alzheimer's disease: many pathways to neuro-degeneration *J. Alz. Dis.* **20** 369–93

[53] Napoli A, Boerakker M J, Tirelli N, Nolte R J, Sommerdijk N A and Hubbell J A 2004 Glucose-oxidase based self-destructing polymeric vesicles *Langmuir* **20** 3487–91

[54] Schafer F Q and Buettner G R 2001 Redox environment of the cell as viewed through the redox state of the glutathione disulfide/glutathione couple *Free Radical Bio. Med.* **30** 1191–212

[55] Duncan R and Kwon G 2005 *Polymeric Drug Delivery Systems* (New York: Marcel Dekker)

[56] Radhakrishnan K, Tripathy J and Raichur A M 2013 Dual enzyme responsive micro-capsules simulating an 'OR' logic gate for biologically triggered drug delivery applications *Chem. Commun.* **49** 5390–2

[57] Wells A and Grandis J R 2003 Phospholipase C-γ1 in tumor progression *Clin. Exp. Metastas.* **20** 285–90

[58] Ferreira A V F 2013 Incorporation of elastase inhibitor in silk fibroin nanoparticles for transdermal delivery *Master's Dissertation* (Portugal: Universidade do Minho)

[59] López-Otín C and Bond J S 2008 Proteases: multifunctional enzymes in life and disease *J. Biol. Chem.* **283** 30433–7

[60] López-Otín C and Hunter T 2010 The regulatory crosstalk between kinases and proteases in cancer *Nat. Rev. Cancer* **10** 278–92

[61] Basel M T, Shrestha T B, Troyer D L and Bossmann S H 2011 Protease-sensitive, polymer-caged liposomes: a method for making highly targeted liposomes using triggered release *ACS Nano* **5** 2162–75

[62] Vicent M J, Greco F, Nicholson R I, Paul A, Griffiths P C and Duncan R 2005 Polymer therapeutics designed for a combination therapy of hormone-dependent cancer *Angew. Chem.* **117** 4129–34

[63] Satchi R, Connors T and Duncan R 2001 PDEPT: polymer-directed enzyme prodrug therapy *Br. J. Cancer* **85** 1070

[64] Duncan R 2005 N-(2-Hydroxypropyl) methacrylamide copolymer conjugates *Polym. Drug Deliv. Syst.* **2005** 1–92

[65] Yoon S-O, Park S-J, Yun C-H and Chung A-S 2003 Roles of matrix metalloproteinases in tumor metastasis and angiogenesis *J. Biochem. Mol. Biol.* **36** 128–37

[66] Crawford H and Matrisian L 1993 Tumor and stromal expression of matrix metalloproteinases and their role in tumor progression *Invas. Metast.* **14** 234–45

[67] Olson E S *et al* 2010 Activatable cell penetrating peptides linked to nanoparticles as dual probes for *in vivo* fluorescence and MR imaging of proteases *Proc. Natl Acad. Sci.* **107** 4311–6

[68] Jiang T, Olson E S, Nguyen Q T, Roy M, Jennings P A and Tsien R Y 2004 Tumor imaging by means of proteolytic activation of cell-penetrating peptides *Proc. Natl Acad. Sci. USA* **101** 17867–72

[69] Imperiale J C, Nejamkin P, del Sole M J, Lanusse C E and Sosnik A 2015 Novel protease inhibitor-loaded nanoparticle-in-microparticle delivery system leads to a dramatic improvement of the oral pharmacokinetics in dogs *Biomaterials* **37** 383–94

[70] Yamashita S I *et al* 1993 Increased expression of membrane-associated phospholipase A2 shows malignant potential of human breast cancer cells *Cancer* **71** 3058–64

[71] Abe T, Sakamoto K, Kamohara H, Yi Hirano, Kuwahara N and Ogawa M 1997 Group II phospholipase A2 is increased in peritoneal and pleural effusions in patients with various types of cancer *Int. J. Cancer* **74** 245–50

[72] Bawarski W E, Chidlowsky E, Bharali D J and Mousa S A 2008 Emerging nano-pharmaceuticals *Nanomedicine* **4** 273–82

[73] Zhang L, Gu F, Chan J, Wang A, Langer R and Farokhzad O 2008 Nanoparticles in medicine: therapeutic applications and developments *Clin. Pharmacol. Ther.* **83** 761–9

[74] Andresen T L, Davidsen J, Begtrup M, Mouritsen O G and Jørgensen K 2004 Enzymatic release of antitumor ether lipids by specific phospholipase A2 activation of liposome-forming prodrugs *J. Med. Chem.* **47** 1694–703

[75] Leiros H K S *et al* 2004 Trypsin specificity as elucidated by LIE calculations, x-ray structures and association constant measurements *Protein Sci.* **13** 1056–70

[76] Xue W, Zhang G and Zhang D 2011 A sensitive colorimetric label-free assay for trypsin and inhibitor screening with gold nanoparticles *Analyst* **136** 3136–41

[77] Radhakrishnan K, Tripathy J, Gnanadhas D P, Chakravortty D and Raichur A M 2014 Dual enzyme responsive and targeted nanocapsules for intracellular delivery of anticancer agents *RSC Adv.* **4** 45961–8

[78] Xu F, Wang W-H, Tan Y-J and Bruening M L 2010 Facile trypsin immobilization in polymeric membranes for rapid, efficient protein digestion *Anal. Chem.* **82** 10045–51

[79] Bernardos A *et al* 2010 Enzyme-responsive intracellular controlled release using nanometric silica mesoporous supports capped with 'saccharides' *ACS Nano* **4** 6353–68

[80] Beith J 1986 Elastases: catalytic and biological properties: biology of extracellular matrix *Regul. Matrix Accumul.* **1** 217–320

[81] Robert L and Hornebeck W 1989 *Elastin and Elastases* (Boca Raton, FL: CRC)

[82] Kim D-H, Chung J H, Son B S, Kim Y J and Lee S G 2014 Effect of a neutrophil elastase inhibitor on ventilator-induced lung injury in rats *J. Thorac. Dis.* **6** 1681

[83] Sun G *et al* 2015 Elevated serum levels of neutrophil elastase in patients with influenza virus-associated encephalopathy *J. Neurol. Sci.* **349** 190–5

[84] Xiong M H, Li Y J, Bao Y, Yang X Z, Hu B and Wang J 2012 Bacteria-responsive multifunctional nanogel for targeted antibiotic delivery *Adv. Mater.* **24** 6175–80

[85] Rao J, Hottinger C and Khan A 2014 Enzyme-triggered cascade reactions and assembly of abiotic block copolymers into micellar nanostructures *J. Am. Chem. Soc.* **2014**

[86] Harnoy A J *et al* 2014 Enzyme-responsive amphiphilic peg-dendron hybrids and their assembly into smart micellar nanocarriers *J. Am. Chem. Soc.* **136** 7531–4

[87] Borgoño C A and Diamandis E P 2004 The emerging roles of human tissue kallikreins in cancer *Nat. Rev. Cancer* **4** 876–90

[88] Yousef G M and Diamandis E P 2001 The new human tissue kallikrein gene family: structure, function, and association to disease 1 *Endocr. Rev.* **22** 184–204

[89] López-Otín C and Matrisian L M 2007 Emerging roles of proteases in tumour suppression *Nat. Rev. Cancer* **7** 800–8

[90] de la Rica R and Velders A H 2011 Supramolecular Au nanoparticle assemblies as optical probes for enzyme-linked immunoassays *Small* **7** 66–9

[91] Wang Z, Qian L, Wang X, Zhu H, Yang F and Yang X 2009 Hollow DNA/PLL microcapsules with tunable degradation property as efficient dual drug delivery vehicles by α-chymotrypsin degradation *Colloid Surface* A **332** 164–1

[92] Gogoi D *et al* 2014 Immobilization of trypsin on plasma prepared Ag/PPAni nanocomposite film for efficient digestion of protein *Mater. Sci. Eng.* C **43** 237–42

[93] Stoytcheva M, Zlatev R, Cosnier S, Arredondo M and Valdez B 2013 High sensitive trypsin activity evaluation applying a nanostructured QCM-sensor *Biosens. Bioelectron.* **41** 862–6

[94] Chuang Y-C *et al* 2010 An optical biosensing platform for proteinase activity using gold nanoparticles *Biomaterials* **31** 6087–95

[95] Gu Z *et al* 2013 Glucose-responsive microgels integrated with enzyme nanocapsules for closed-loop insulin delivery *ACS Nano* **7** 6758–66

[96] Lee D *et al* 2015 Establishment of controlled insulin delivery system using glucose-responsive double-layered nanogel *RSC Adv.* **5** 14482–91

[97] de la Rica R, Fratila R M, Szarpak A, Huskens J and Velders A H 2011 Multivalent nanoparticle networks as ultrasensitive enzyme sensors *Angew. Chem.* **123** 5822–5

Smart Internal Stimulus-Responsive Nanocarriers
for Drug and Gene Delivery

Mahdi Karimi, Parham Sahandi Zangabad, Amir Ghasemi and Michael R Hamblin

Chapter 4

Redox-responsive micro/nanocarriers

4.1 Redox-responsive nano drug/gene delivery systems

In recent decades, the rate of research on redox-responsive systems has increased considerably, mainly because of the abundance of redox-active stimuli that can be utilized to develop a smart DGDS [1, 2]. The advantages of this group of stimuli are:

- glutathione disulfide/reduced gluathione (GSSG/GSH) is a very important redox couple in mammalian cells and has been utilized in reduction-responsive DDS [3–9];
- redox-responsive delivery vehicles are simple to construct;
- the activation procedure is simple in chemical terms.

The tripeptide GSH is an excellent internal stimulus for efficient drug/gene release. The GSH concentration gradient plays a significant role in redox-responsive nano-DGDS [10]. For instance, the normal intracellular concentration of GSH falls within a wide range (2–10 mM). By contrast, the GSH concentration in extracellular fluid is only about 2–20 μM [11]. When DDS are used for gene delivery it is important to note that free DNA fragments are rapidly degraded in blood by serum nucleases, and it is therefore necessary for exogenous DNA to be encapsulated for delivery in order for it to survive long enough to penetrate into the target cells where it will be released [12]. In this section we focus on different GSH-responsive nanocarriers (e.g. nanogels, polymerosomes, capsules and micelles) that meet the requirements for a smart DGDS activated by a redox stimulus.

4.2 Nanogels

Nanogels (NP composed of a crosslinked hydrophilic polymer network [13]) are biocompatible three-dimensional materials that can be used for the encapsulation of cargos such as anticancer drugs, plasmid DNA and imaging probes [14]. Nanogels

4-1

have a high water content and their size ranges from tens of nanometers to submicrons [14, 15]. In this group of materials, macromolecules can be protected from the environment until they undergo triggered release by being located within the pores of nanogels.

Matyjaszewski *et al* [16, 17] synthesized a reduction-sensitive nanogel in which they utilized the disulphide–thiol exchange reaction and the process of inverse mini-emulsion atom transfer radical polymerization (ATRP). These novel nanogels can be loaded with a wide range of water-soluble biomolecules, ranging from anticancer drugs through proteins to carbohydrates [16, 17]. For example, nanogels were loaded with DOX with 50–70% efficiency and the disulfide crosslinked nanovehicles were non-toxic, but after the addition of 20 wt% GSH they inhibited HeLa cell growth significantly. Caruso *et al* recently reported a method for preparing reduction-sensitive DOX-loaded PEGylated nanoporous polymer-spheres (NPSPEG–DOX). First, they loaded and immobilized alkyne- or azide-functionalized PEG into a MSN template via click chemistry; then they carried out crosslinking of PEG and covalent binding of DOX through biocompatible linkers containing disulfide bonds; and finally they dissolved away the MSN templates [18]. Under reductive conditions (5 mM GSH), the nanospheres disassembled to release DOX. Biocompatible and degradable nanogels have also been prepared through covalent binding of thiol-functionalized star-shaped poly(ethyleneoxide-co-propylene oxide) with linear polyglycidol in an inverse mini-emulsion via the formation of disulfide bonds [19]. These nanogels were degraded after 6 h incubation in 10 mM GSH. Finally, the temperature, pH and reduction triple-responsive nanogels were produced through mini-emulsion copolymerization of monomethyloligo(ethylene glycol) acrylate (OEGA) and an orthoester-containing acrylic monomer, 2-(5,5-dimethyl-1,3-dioxan-2-yloxy) ethyl acrylate (DMDEA), using bis(2-acryloyloxyethyl) disulfide (BADS) as a crosslinker [20].

The redox-responsive properties of the nanogels are strongly dependent on the degree of crosslinking of the polymer and its composition. For instance, a novel redox-responsive nanogel was produced by utilizing mini-emulsion radical-mediated polymerization of monomethyl OEGA, with BADS as a stabilizer, and an orthoester-containing acrylic monomer, DMDEA [20]. The results showed that the synthesized nanogels, with a high degree of crosslinking, had good cytotoxicity towards tumor cells through the delivery of PTX, a hydrophobic drug, at a pH of about 7.4. In another similar study, Nguyen *et al* [21] prepared heparin–pluronic nanogels with a mean diameter of 115.7 nm containing a redox-sensitive disulphide bond, using the thiol-reactive vinylsulfone (VS) group. It was proposed that these nanogels be employed as multifunctional nanogels for intracellular protein delivery and release.

4.3 Polymersomes

Polymersomes have attracted the attention of many researchers for applications in biotechnology (in particular for intracellular protein delivery) because of their ability to mimic the properties of viruses and cell structures [22]. They can also dissolve drugs without aggregation [23–25]. Hubell *et al* [26–28] synthesized polymersomes as a smart DDS that can be triggered at sites of inflammation and in endo-/lysosomal compartments. Their results showed that these polymersomes can be used in either oxidation- [27–29] or reduction-responsive [26] DDS. For instance, for reduction-responsive conditions, the PPS block contains hydrophobic thioether moieties that stabilize the vesicular bilayer. However, in the oxidation-responsive condition, the polymersomes are formed through a developed self-assembly system, with a triblock copolymer connected to two hydrophilic blocks of PEG around a hydrophobic block of poly(propylene sulfide) (PPS): PEG–PPS–PEG. The functionalized groups are oxidized by exposure to an oxidizing agent (e.g. hydrogen peroxide) and can produce hydrophilic sulfoxide and sulfone functionalities. The researchers concluded that this loss of hydrophobic character destabilizes the vesicle by rupturing the lamellar bilayer [27–29].

4.4 Nanocapsules

Nanocapsules are another class of NP that have been utilized in the experimental and clinical delivery of therapeutic and diagnostic agents [30]. The specific properties of nanocapsules allow them to penetrate through the membrane of the cell to release their cargos. Their structure comprises an active material (core) surrounded by a protective matrix (shell), allowing release through simple oxidation of the protective coating. The surface modification of these vehicles can be tailored to allow them to perform several different functions simultaneously [30–32].

In general, three different physico-chemical mechanisms can be applied to release the cargo from nanocapsule polymeric shells [33]. Hollow capsules can be highly versatile vehicles, capable of being used for the encapsulation and controlled delivery of diverse bioactive molecules, including drugs, nucleic acids, peptides and proteins [34, 35].

A new reduction-degradable capsule based on layer-by-layer (LBL) assembly was developed. Zelikin *et al* [36, 37] assembled poly(vinylpyrrolidone) (PVPON) and thiolated poly(methacrylic acid) (PMASH) into silica gels. In this system, the silica-based core was degraded through the generation of disulfide bonds from thiol groups in PMASH. As a consequence, the polymer chains were destroyed in the physiological buffer and PVPON was then released. These capsules were stable in oxidizing conditions. However, they were disassembled in a reducing environment (e.g. inside living cells) so as to release the encapsulated drug.

Kim *et al* [38] recently reported on a new template-free method for synthesizing reduction-responsive polymer nanocapsules based on the self-assembly of amphiphilic cucurbit[6]uril (CB[6]) followed by shell-crosslinking with a disulfide-containing crosslinker. The average diameter of the resulting capsules was approximately 70 nm and they had a hollow interior, surrounded by an approximately 2 nm thick shell. The nanocapsules aggregated and collapsed after 30 min treatment with DTT. *In vitro* release studies demonstrated that the encapsulated carboxyfluorescein (CF) was released quickly in response to 100 mM DTT. The galactose-coated capsules were taken up by HepG2 cells and showed the burst release kinetics of CF inside the cells. Zhang *et al* [29] reported on the preparation of reduction-sensitive hollow polyelectrolyte nanocapsules made from dextran sulfate and chitosan linked-cysteamine through LBL adsorption on cyclodextrin-coated nanospheres of silica, which were then thiol crosslinked to allow the silica core to be removed.

The copolymerization of monomers with crosslinkers that form disulfide linkages is another method for synthesizing polymeric nanocapsules, as reported by Yan *et al* [40, 41]. Yet another strategy is the post-crosslinking of a polymer that has dithiols into disulfide bonds, as reported by Huang *et al* [42]. Silica NP with disulfide bonds were evaluated recently for controlled release in biomedical applications [43, 44].

4.5 Micelles

4.5.1 Drug delivery from micelles

Impressive progress in bioengineering has opened the door to the study of a new class of materials, micelles, as a useful carrier for smart drug delivery. Micelles have advantages such as controllability of drug release, easy preparation and stability, while hydrophobic drugs can be loaded into their core [45].

There are many different classes of DDS, including polymeric micelles, nano-particle drug carriers, dendrimers, lipoprotein-based drug carriers, liposomes, etc. Micelles are biodegradable, non-toxic, biocompatible and non-immunogenic—all desirable features for an ideal DDS [46], which must also stay unrecognized by the defense mechanism of the host [47].

4.5.2 Polymeric micelles

The small size (diameter = 10–100 nm) of polymeric micelles makes them an attractive option for the preparation of a GSH-responsive carrier. Some of the advantages of this material are:

- high structural stability;
- low toxicity;
- high water solubility;
- large capacity for drug loading;
- incorporation of various chemical species.

There are also some downsides, such as slow extravasation, immature drug-incorporation technology, difficult polymer synthesis and possible chronic liver toxicity due to slow metabolism.

Polymeric micelles are generated by the spontaneous self-assembly of amphiphilic block copolymers above the minimum micellar concentration [48]. This critical micellar concentration (CMC) is the concentration of surfactants above which micelles form, and all additional surfactant in the system goes into the micelles [49].

4.5.3 Redox-sensitive polymeric micelles

De-crosslinking and disassembly or full destabilization of redox-sensitive micelles may be caused by the action of intracellular GSH that comes with the reduction of disulfide bonds in polymeric assemblies [50–53]. Anton *et al* [54] reported that the reversible redox reactions occurring in a bio-metallic polymer composite containing viologen or ferrocene can change the charge density and thus the solubility. In another study, it was concluded that the oxidation of redox-active micelles containing a hydrophobic ferrocenyl-alkyl moiety in the block copolymer shifted the hydrophilic/hydrophobic balance; as a result, the micelles fragmented into water-soluble monomers. Takeoka *et al* [55] showed that the hydrophobic model drug released from these micelles could be controlled by selective electrochemical oxidation of the ferrocenyl alkyl moiety, with zero-order kinetics.

Micelles with detachable shells

Recently, researchers have focused on the inhibition of cancer cell growth (such as A549 lung cancer) after pre-treatment with glutathione monoester (GSH-OEt). GSH-OEt increases the intracellular GSH level in an artificial fashion. This approach may be used to overcome the MDR of cancer cells. Wang *et al* [56, 57] synthesized a reduction-sensitive shell-detachable micelle. Their results demonstrated that the disulfide-linked copolymer of PCL and the hydrophilic poly(ethyl ethylene phosphate) (PEEP) (PCL–SS–PEEP) releases DOX from shell-detachable micelles.

A bioreducible amphiphilic triblock-polymerosome with a diameter of 256 nm was described recently. At <7.4 pH it released a drug to SCC cells and the release increased in the presence of 10 mm GSH [5]. Yoo and Park [58] reported GSH-triggered drug release from camptothecin (CPT)-loaded PEG–SS-poly(g-benzyl L-glutamate) (PEG–SS–PBLG) micelles, resulting in higher toxicity towards SCC7 cancer cells compared with CPT loaded into PEG–b–PBLG micelles (reduction-non-sensitive control). Yuan *et al* determined that intracellular GSH concentration could affect gold NP loaded with a micro-RNA (miR-122) that targeted the Bcl-W pathway and resulted in apoptosis of liver cancer cells (Hep G2). Conjugation of FA also helped to target the cancer cells [59]. In another study, the authors tested the attachment of CPT to a MSN containing disulfide bridges formed from a mercapto-functionalized silica hybrid that can release a drug to kill HeLa cells under the influence of GSH [60].

Li *et al* [61] and Shi *et al* [62] reported that reduction-sensitive PEG–SS–poly (e-benzyloxycarbonyl-L-lysine) (mPEG-SS-PzLL) and PEG-SS-polyleucine (PEG–SS–Pleu) micelles showed enhanced DOX release in response to 10 mM DTT. Li *et al* [63] prepared shell-detachable micelles based on a six-armed star PCL–SS–PEG, which demonstrated accelerated release of DOX in the presence of 10 mM DTT. DOX-loaded micelles inhibited the growth of MCF-7 cells, depending on GSH. Yan *et al* [64] constructed shell-detachable micelles from the amphiphilic hyperbranched multi-armed star PLA–SS–poly(2-ethoxy-2-oxo-1,3,2-dioxaphospholane) (PEP) copolymer, and Oh *et al* [65] also reported on the synthesis and reduction-triggered shell-shedding of PEG–SS–PLA micelles.

Micelles with reduction-sensitive cores

Fan *et al* [66] prepared micelles that degraded in the presence of reducing thiols, using amphiphilic grafted copolymers of hydrophobic poly(amido amine) (SS–PAA) and PEG (SS–PAA–gPEG) linked by disulfide bonds. The DOX was almost quantitatively released *in vitro* after 10 h treatment with 1 mM DTT, while only ~25% of the DOX was released in 24 h in the absence of DTT. The IC50 of the DOX-loaded SS–PAA–g–PEG micelles was determined to be 0.0647 mg mL^{-1} for HepG2 cells and 0.0494 mg mL^{-1} for HeLa cells; these values were only slightly higher than the IC50 of free DOX.

Novel reduction-sensitive amphiphilic hyperbranched polyphosphates (HPHDP) have been prepared through the self-condensation and ring-opening polymerization (SCROP) of 2-[(2-hydroxyethyl)-disulfanyl]ethoxy-2-oxo-1,3,2-dioxaphospholane (PHDP) [67]. These materials can form micelles with a small diameter and a multi-core/shell structure. In a subsequent study [68], micelles were prepared from amphiphilic hyper-branched block co-polyphosphates with a reduction-sensitive hydrophobic core and a hydrophilic shell, with the intention of killing HeLa cells pre-treated with GSH-OEt.

GSH-sensitive micelles can be also prepared from amphiphilic copolymers containing disulfide bonds in the hydrophobic segments, which can be cleaved in response to elevated GSH concentrations, leading to the disassembly of the micelles and concomitant drug release [69]. Although drug release was relatively slow even in the presence of 70 mM GSH, the cytotoxicity of DOX-loaded micelles was positively correlated with the intra-cellular GSH level in MCF-7 cells. Reduction-sensitive micelles based on poly(ethylene oxide)–b–poly(N-methacryloyl-N0-(t-butyloxycarbonyl)cystamine) (PEO–b–PMABC) diblock copolymers have also shown faster DOX release and higher anticancer efficacy than the reduction-insensitive controls [70]. An amphiphilic copolymer micelle that contained diselenide blocks was designed to be responsive to both reduction and oxidation. It underwent fast disassembly at a low concentration of GSH (only 0.01 mg mL^{-1}) and also with a low concentration of the oxidant H_2O_2 (0.01%) [71].

Another micellar preparation was made from similar polymers, containing selenium, poly(ethylene oxide-b-acrylic acid) block copolymers (PEO–b–PAA–Se). These block copolymers self-assembled in aqueous solution and formed spherical micellar aggregates. The selenide group of PEO–b–PAA–Se was oxidized to hydrophilic selenoxide with 0.1% hydrogen peroxide, leading to disassembly. The fluorescent dye (Nile red) showed fast release upon the addition of 0.1% hydrogen peroxide. The oxidation state of selenoxide could be reversed back to selenide after reduction with Vitamin C [72].

Reduction-sensitive core-crosslinked micelles

Being stable *in vivo* and thus avoiding premature drug leakage after intravenous administration is a practical challenge for micellar carriers [73]. Studies have demonstrated that effective crosslinking of micelles can overcome the instability problem [74]. However, it should be noted that because the release of drugs may be difficult once the micelles arrive at the target sites, which could result in low drug delivery, the micelles should not be too stable, either. The use of disulfide crosslinks that are reversible under intracellular conditions is a good strategy to solve the stability/drug release dilemma for micelles. Using divalent metal cations (Ca^{2+}) as a template and crosslinking the ionic cores with cystamine, Bronich *et al* [75] prepared poly(ethylene oxide)-b-poly(methacrylic acid) (PEO–b–PMAc) micelles. Interestingly, these micelles demonstrated a high level of DOX loading (50% w/w). *In vitro* release studies showed a significant acceleration of DOX release from cystamine-crosslinked micelles in the presence of GSH or cysteine in the media. 75% of DOX was released in 1 h in response to 10 mM GSH. Stenzel *et al* [76] produced stable nucleoside-containing block copolymer micelles through sequential reversible addition–fragmentation chain transfer (RAFT) copolymerization of polyethylene glycol methyl ether methacrylate, 5-O-methacryloyl-uridine and bis(2-methacry-loyloxyethyl)disulfide (DSDMA, a bioreducible crosslinker). In the presence of 0.65 mM DTT, the core-crosslinked (CCL) micelles hydrolyzed into free block copolymers in less than 60 min. As expected, CCL micelles showed a rather slow release of riboflavin. By contrast, the addition of 0.65 mM DTT led to fast drug release, with a pattern akin to that of the non-crosslinked control (about 60–70% release in 7 h). Liu *et al* [77] also used RAFT polymerization to prepare two types of degradable thermo-responsive CCL micelles.

References

[1] Chuan X *et al* 2014 Novel free-paclitaxel-loaded redox-responsive nanoparticles based on a disulfide-linked poly(ethylene glycol)-drug conjugate for intracellular drug delivery: synthesis, characterization and antitumor activity *in vitro* and *in vivo Mol. Pharm.* **11** 3656–70

[2] Phillips D J and Gibson M I 2013 Redox-sensitive materials for drug delivery: targeting the correct intracellular environment, tuning release rates, and appropriate predictive systems *Antioxid. Redox Signal.* **21** 786–803

[3] Zhang J, Yang F, Shen H and Wu D 2012 Controlled formation of microgels/nanogels from a disulfide-linked core/shell hyperbranched polymer *ACS Macro Lett.* **1** 1295–9

[4] Yuan W, Zou H, Guo W, Shen T and Ren J 2013 Supramolecular micelles with dual temperature and redox responses for multi-controlled drug release *Polymer Chem.* **4** 2658–61

[5] Thambi T, Deepagan V, Ko H, Lee D S and Park J H 2012 Bioreducible polymersomes for intracellular dual-drug delivery *Journal of Materials Chemistry* **22** 22028–36

[6] Han D, Tong X and Zhao Y 2012 Block copolymer micelles with a dual-stimuli-responsive core for fast or slow degradation *Langmuir* **28** 2327–31

[7] Yu Z-Q, Sun J-T, Pan C-Y and Hong C-Y 2012 Bioreducible nanogels/microgels easily prepared via temperature induced self-assembly and self-crosslinking *Chem. Commun.* **48** 5623–5

[8] Yan Y, Wang Y, Heath J K, Nice E C and Caruso F 2011 Cellular association and cargo release of redox-responsive polymer capsules mediated by exofacial thiols *Adv. Mater.* **23** 3916–21

[9] Aleksanian S, Khorsand B, Schmidt R and Oh J K 2012 Rapidly thiol-responsive degradable block copolymer nanocarriers with facile bioconjugation *Polymer Chem.* **3** 2138–47

[10] Schafer F Q and Buettner G R 2001 Redox environment of the cell as viewed through the redox state of the glutathione disulfide/glutathione couple *Free Radical Bio. Med.* **30** 1191–212

[11] Cheng R, Feng F, Meng F, Deng C, Feijen J and Zhong Z 2011 Glutathione-responsive nano-vehicles as a promising platform for targeted intracellular drug and gene delivery *J. Controlled Release* **152** 2–12

[12] Mintzer M A and Simanek E E 2008 Nonviral vectors for gene delivery *Chem. Rev.* **109** 259–302

[13] Sultana F, Manirujjaman M, Imran-Ul-Haque M A and Sharmin S 2013 An overview of nanogel drug delivery system *J. Appl. Pharmaceut. Sci.* **3** S95–S105

[14] Kabanov A V and Vinogradov S V 2009 Nanogels as pharmaceutical carriers: finite networks of infinite capabilities *Angew. Chem. Int. Ed.* **48** 5418–29

[15] Oh J K, Drumright R, Siegwart D J and Matyjaszewski K 2008 The development of microgels/nanogels for drug delivery applications *Prog. Polym. Sci.* **33** 448–77

[16] Oh J K *et al* 2007 Biodegradable nanogels prepared by atom transfer radical polymerization as potential drug delivery carriers: synthesis, biodegradation, *in vitro* release and bioconjugation *J. Am. Chem. Soc.* **129** 5939–45

[17] Oh J K, Siegwart D J and Matyjaszewski K 2007 Synthesis and biodegradation of nanogels as delivery carriers for carbohydrate drugs *Biomacromolecules* **8** 3326–31

[18] Yap H P, Johnston A P, Such G K, Yan Y and Caruso F 2009 Click-engineered, bioresponsive, drug-loaded peg spheres *Adv. Mater.* **21** 4348–52

[19] Groll J, Singh S, Albrecht K and Moeller M 2009 Biocompatible and degradable nanogels via oxidation reactions of synthetic thiomers in inverse miniemulsion *J. Polym. Sci. A: Polym. Chem.* **47** 5543–9

[20] Qiao Z-Y, Zhang R, Du F-S, Liang D-H and Li Z-C 2011 Multi-responsive nanogels containing motifs of ortho ester, oligo (ethylene glycol) and disulfide linkage as carriers of hydrophobic anti-cancer drugs *J. Controlled Release* **152** 57–66

[21] Nguyen D H, Joung Y K, Choi J H, Moon H T and Park K D 2011 Targeting ligand-functionalized and redox-sensitive heparin–Pluronic nanogels for intracellular protein delivery *Biomed. Mater.* **6** 055004

[22] Liu G *et al* 2010 The highly efficient delivery of exogenous proteins into cells mediated by biodegradable chimaeric polymersomes *Biomaterials* **31** 7575–85

[23] Meng F and Zhong Z 2011 Polymersomes spanning from nano to microscales: advanced vehicles for controlled drug delivery and robust vesicles for virus and cell mimicking *J. Phys. Chem. Lett.* **2** 1533–9

[24] LoPresti C, Lomas H, Massignani M, Smart T and Battaglia G 2009 Polymersomes: nature inspired nanometer sized compartments *J. Mater. Chem.* **19** 3576–90

[25] Discher D E and Ahmed F 2006 Polymersomes *Annu. Rev. Biomed. Eng.* **8** 323–41

[26] Cerritelli S, Velluto D and Hubbell J A 2007 PEG–SS–PPS: reduction-sensitive disulfide block copolymer vesicles for intracellular drug delivery *Biomacromolecules* **8** 1966–72

[27] Napoli A, Valentini M, Tirelli N, Müller M and Hubbell J A 2004 Oxidation-responsive polymeric vesicles *Nat. Mater.* **3** 183–9

[28] Napoli A, Boerakker M J, Tirelli N, Nolte R J, Sommerdijk N A and Hubbell J A 2004 Glucose-oxidase based self-destructing polymeric vesicles *Langmuir* **20** 3487–91

[29] Zhang H *et al* 2010 Oxidizing-responsive vesicles made from 'tadpole-like supramolecular amphiphiles' based on inclusion complexes between driving molecules and β-cyclodextrin *Colloids Surf. Physicochem. Eng. Aspects* **363** 78–85

[30] Torchilin V P 2008 *Multifunctional Pharmaceutical Nanocarriers* (Berlin: Springer)

[31] Avramoff A, Laor A, Kitzes-Cohen R, Farin D and Domb A 2007 Comparative *in vivo* bioequivalence and *in vitro* dissolution of two cyclosporin A soft gelatin capsule formulations *Int. J. Clin. Pharmacol. Ther.* **45** 126–32

[32] Thassu D, Pathak Y and Deleers M 2007 Nanoparticulate drug-delivery systems: an overview *Drugs Pharmaceut. Sci.* **166** 1

[33] Nagavarma B, Hemant K Y, Ayaz A, Vasudha L and Shivakumar H 2012 Different techniques for preparation of polymeric nanoparticles—a review *Asian J. Pharmaceut. Clin. Res.* **5** 16–23

[34] Sukhorukov G B *et al* 2005 Nanoengineered polymer capsules: tools for detection, controlled delivery and site-specific manipulation *Small* **1** 194–200

[35] De Geest B G, Sanders N N, Sukhorukov G B, Demeester J and De Smedt S C 2007 Release mechanisms for polyelectrolyte capsules *Chem. Soc. Rev.* **36** 636–49

[36] Zelikin A N, Li Q and Caruso F 2008 Disulfide-stabilized poly (methacrylic acid) capsules: formation, cross-linking and degradation behavior *Chem. Mater.* **20** 2655–61

[37] Zelikin A N, Quinn J F and Caruso F 2006 Disulfide cross-linked polymer capsules: en route to biodeconstructible systems *Biomacromolecules* **7** 27–30

[38] Kim E *et al* 2010 Facile, template-free synthesis of stimuli-responsive polymer nanocapsules for targeted drug delivery *Angew. Chem.* **122** 4507–10

[39] Shu S, Zhang X, Wu Z, Wang Z and Li C 2010 Gradient cross-linked biodegradable polyelectrolyte nanocapsules for intracellular protein drug delivery *Biomaterials* **31** 6039–49

[40] Yan L, Wang Z-K, Yan J-J, Han L-F, Zhou Q-H and You Y-Z 2013 Selectively grafting polymer from the interior and/or exterior surfaces of bioreducible and temperature-responsive nanocapsules *Polymer Chem.* **4** 1243–9

[41] Zhao M *et al* 2011 Redox-responsive nanocapsules for intracellular protein delivery *Biomaterials* **32** 5223–30

[42] Huang X, Appelhans D, Formanek P, Simon F and Voit B 2012 Tailored synthesis of intelligent polymer nanocapsules: an investigation of controlled permeability and pH-dependent degradability *ACS Nano* **6** 9718–26

[43] Piao Y, Burns A, Kim J, Wiesner U and Hyeon T 2008 Designed fabrication of silica-based nanostructured particle systems for nanomedicine applications *Adv. Funct. Mater.* **18** 3745–58

[44] Trewyn B G, Slowing I I, Giri S, Chen H-T and Lin V S-Y 2007 Synthesis and functionalization of a mesoporous silica nanoparticle based on the sol–gel process and applications in controlled release *Acc. Chem. Res.* **40** 846–53

[45] Alvarez-Lorenzo C and Concheiro A 2013 *Smart Mater. Drug Deliv.* vol 1 (Cambridge: Royal Society of Chemistry)

[46] Scott R C, Crabbe D, Krynska B, Ansari R and Kiani M F 2008 Aiming for the heart: targeted delivery of drugs to diseased cardiac tissue **5** 459–70

[47] Bertrand N and Leroux J-C 2012 The journey of a drug-carrier in the body: an anatomo-physiological perspective *J. Controlled Release* **161** 152–63

[48] Kataoka K, Harada A and Nagasaki Y 2001 Block copolymer micelles for drug delivery: design, characterization and biological significance *Adv. Drug Deliv. Rev.* **47** 113–31

[49] McNaught A D and McNaught A D 1997 *Compendium of Chemical Terminology* (Oxford: Blackwell)

[50] Li Y, Lokitz B S, Armes S P and McCormick C L 2006 Synthesis of reversible shell cross-linked micelles for controlled release of bioactive agents *Macromolecules* **39** 2726–8

[51] Kakizawa Y, Harada A and Kataoka K 2001 Glutathione-sensitive stabilization of block copolymer micelles composed of antisense DNA and thiolated poly (ethylene glycol)-b lock-poly (l-lysine): a potential carrier for systemic delivery of antisense DNA *Biomacromolecules* **2** 491–7

[52] Kakizawa Y, Harada A and Kataoka K 1999 Environment-sensitive stabilization of core–shell structured polyion complex micelle by reversible cross-linking of the core through disulfide bond *J. Am. Chem. Soc.* **121** 11247–8

[53] Ghosh S, Basu S and Thayumanavan S 2006 Simultaneous and reversible functionalization of copolymers for biological applications *Macromolecules* **39** 5595–7

[54] Anton P, Heinze J and Laschewsky A 1993 Redox-active monomeric and polymeric surfactants *Langmuir* **9** 77–85

[55] Takeoka Y *et al* 1995 Electrochemical control of drug release from redox-active micelles *J. Controlled Release* **33** 79–87

[56] Wang Y-C, Wang F, Sun T-M and Wang J 2011 Redox-responsive nanoparticles from the single disulfide bond-bridged block copolymer as drug carriers for overcoming multidrug resistance in cancer cells *Bioconj. Chem.* **22** 1939–45

[57] Tang L-Y, Wang Y-C, Li Y, Du J-Z and Wang J 2009 Shell-detachable micelles based on disulfide-linked block copolymer as potential carrier for intracellular drug delivery *Bioconj. Chem.* **20** 1095–9

[58] Thambi T, Yoon H Y, Kim K, Kwon I C, Yoo C K and Park J H 2011 Bioreducible block copolymers based on poly (ethylene glycol) and poly (γ-benzyl L-glutamate) for intracellular delivery of camptothecin *Bioconj. Chem.* **22** 1924–31

[59] Yuan Y, Zhang X, Zeng X, Liu B, Hu F and Zhang G 2014 Glutathione-mediated release of functional MiR-122 from gold nanoparticles for targeted induction of apoptosis in cancer treatment *J. Nanosci. Nanotechnol.* **14** 5620–7

[60] Botella P, Muniesa C, Vicente V, Fabregat K and Cabrera A 2014 Controlled intracellular release of camptothecin by glutathione-driven mechanism *Nanotechnology* vol 2 (Boca Raton, FL: Taylor & Francis)

[61] Ren T-B, Xia W-J, Dong H-Q and Li Y-Y 2011 Sheddable micelles based on disulfide-linked hybrid PEG-polypeptide copolymer for intracellular drug delivery *Polymer* **52** 3580–6

[62] Wen H-Y *et al* 2011 Rapidly disassembling nanomicelles with disulfide-linked PEG shells for glutathione-mediated intracellular drug delivery *Chem. Commun.* **47** 3550–2

[63] Ren T-B, Feng Y, Zhang Z-H, Li L and Li Y-Y 2011 Shell-sheddable micelles based on star-shaped poly (ε-caprolactone)–SS–poly (ethyl glycol) copolymer for intracellular drug release *Soft Matter* **7** 2329–31

[64] Liu J *et al* 2011 Bioreducible micelles self-assembled from amphiphilic hyperbranched multiarm copolymer for glutathione-mediated intracellular drug delivery *Biomacromolecules* **12** 1567–77

[65] Khorsand Sourkohi B, Cunningham A, Zhang Q and Oh J K 2011 Biodegradable block copolymer micelles with thiol-responsive sheddable coronas *Biomacromolecules* **12** 3819–25

[66] Sun Y *et al* 2010 Disassemblable micelles based on reduction-degradable amphiphilic graft copolymers for intracellular delivery of doxorubicin *Biomaterials* **31** 7124–31

[67] Liu J *et al* 2011 Molecular self-assembly of a homopolymer: an alternative to fabricate drug-delivery platforms for cancer therapy *Angew. Chem.* **123** 9328–32

[68] Liu J *et al* 2011 Redox-responsive polyphosphate nanosized assemblies: a smart drug delivery platform for cancer therapy *Biomacromolecules* **12** 2407–15

[69] Ryu J-H, Roy R, Ventura J and Thayumanavan S 2010 Redox-sensitive disassembly of amphiphilic copolymer based micelles *Langmuir* **26** 7086–92

[70] Sun P, Zhou D and Gan Z 2011 Novel reduction-sensitive micelles for triggered intracellular drug release *J. Controlled Release* **155** 96–103

[71] Ma N, Li Y, Xu H, Wang Z and Zhang X 2009 Dual redox responsive assemblies formed from diselenide block copolymers *J. Am. Chem. Soc.* **132** 442–3

[72] Ren H, Wu Y, Ma N, Xu H and Zhang X 2012 Side-chain selenium-containing amphiphilic block copolymers: redox-controlled self-assembly and disassembly *Soft Matter* **8** 1460–6

[73] Bae Y H and Yin H 2008 Stability issues of polymeric micelles *J. Controlled Release* **131** 2–4

[74] O'Reilly R K, Hawker C J and Wooley K L 2006 Cross-linked block copolymer micelles: functional nanostructures of great potential and versatility *Chem Soc. Rev.* **35** 1068–83

[75] Kim J O, Sahay G, Kabanov A V and Bronich T K 2010 Polymeric micelles with ionic cores containing biodegradable cross-links for delivery of chemotherapeutic agents *Biomacromolecules* **11** 919–26

[76] Zhang L, Liu W, Lin L, Chen D and Stenzel M H 2008 Degradable disulfide core-cross-linked micelles as a drug delivery system prepared from vinyl functionalized nucleosides via the RAFT process *Biomacromolecules* **9** 3321–31

[77] Zhang J, Jiang X, Zhang Y, Li Y and Liu S 2007 Facile fabrication of reversible core cross-linked micelles possessing thermosensitive swellability *Macromolecules* **40** 9125–32

Mahdi Karimi, Parham Sahandi Zangabad, Amir Ghasemi and Michael R Hamblin

Chapter 5

Biomolecule-sensitive nanocarriers

5.1 Introduction

In biological systems, specific ions or biological molecules such as hormones can be released that are associated with the functions of a natural feedback system. By mimicking such natural feedback systems, these functions can be used in smart DGDS [1]. In chapter 3, which covered the major role played by enzymes in targeted delivery systems, enzyme-responsive DGDS were discussed exhaustively. In this section we cover systems that can respond to other (non-enzymic) biomolecules, which can then induce targeted delivery in biological environments.

5.2 Adenosine-5′-triphosphate-responsive

Adenosine-5′-triphosphate (ATP) is a multifunctional nucleotide and is responsible for the production and degradation of several cellular compounds (as a coenzyme), proper cell function membrane transport and muscle contraction, while it is the main source of cellular energy for metabolism and signaling [2, 3]. The concentration of ATP in the intracellular environment (1–20 mM) is much higher than in the extracellular environment (less than 5 μM). This distinct difference in ATP concentration can be used for the production of ATP-triggered DDS. Therefore, in recent years many studies have been conducted in this area (mostly by Mo *et al* [2–6]). For example, in a recent study an ATP-responsive drug nanocarrier was synthesized to carry DOX. It was composed of GO assembled into nanoaggregates and crosslinked through ATP-sensitive DNA strands (i.e. single-stranded DNA (ssDNA)). This system showed high loading efficiency, controlled drug release and site-specific targeting [5]. In another study by Mo *et al* [6], they reported that an ATP-binding aptamer incorporating a DNA motif functionalizing a polymeric nanogel nanovehicle was able to recognize ATP (a metabolic trigger) precisely and facilitate controlled drug release depending on the ATP concentration. The final

doi:10.1088/978-1-6817-4257-1ch5

nanogel formulation had three compartments, including the ATP-sensitive DNA motif with loaded drugs, protamine and a hyaluronic acid (HA) ligand crosslinked shell. The ATP-mediated nanocarrier was equipped with the crosslinked shell (as a tumor targeting moiety). The DNA scaffold comprised its complementary single-stranded DNA (cDNA) and an ATP-recognizing aptamer (as an ATP detector). The positively charged protamine was used for compression of the DNA scaffold with many negative charges into a positively charged core complex, resulting in cell penetration, endosomal escape and nuclear targeting. A protective shell was formed by coating the negatively charged HA onto the core complex, which provided targeting to tumor sites through binding ligands to the receptors on the surface of cancer cells. DOX was loaded by intercalating between the GC pairs of the DNA motif. The final nanogel structures were formed by photo-crosslinking activated by UV irradiation. Thereafter, the nanogels were intravenously injected and accumulated at tumor sites through passive and active targeting. The HA shell was degraded in hyaluronidase (HAase)-rich environments such as tumor cellular endocytic vehicles (including endosomes and lysosomes (endolysosomes)), leading to the exposure of the cationic complex of protamine with DOX-intercalated DNA duplex and facilitating the intracellular delivery. The protamine enhanced the endosomal escape of the complex and thus effective transport of DOX/duplex complex into the cytosol was achieved. The aptamer tertiary structure was stabilized by exposure to the ATP-rich milieu of the cytosol and the binding of its target ATP, which induced dissociation of the compressed DNA-based DOX/duplex. This structural change of duplex-to-aptamer resulted in the liberation of cDNA and the subsequent specific release of intercalated DOX from the nanovehicles via a conformational switch. Thus cytotoxicity was enhanced through DOX accumulation in the nuclei, leading to apoptosis (figure 5.1(a)). In another study, He *et al* [2] designed ATP-triggered MSN functionalized with an aptamer as a capping agent. Here, one arm bore single-stranded DNA1 (arm ssDNA1) and another arm bore single-stranded DNA2 (arm ssDNA2), and these were used to hybridize the ATP aptamer and form a sandwich-type DNA structure which was subsequently grafted onto the surface of the MSN. The guest molecules were then encapsulated into the pores of the MSN. Therefore, the pores of the MSN were blocked, leading to the release of the guest drug molecules (e.g. $Ru(bipy)_3^{2+}$ molecules) being impeded. Through exposure to an ATP-rich environment, competitive binding between ATP and the ATP aptamer occurred, resulting in the removal of the ATP molecules from the pores and allowing the two arms (ssDNA1 and ssDNA2) to separate. The release of the guest molecules was possible because of the flexibility of the ssDNA moieties (figure 5.1(b)). The ATP-sensitive behavior results indicated high selectivity towards ATP analogues, good capping efficiency, high loading efficiency (215.0 μmol g^{-1} SiO$_2$) and desirable release behavior in the presence of ATP, in addition to high stability in mouse serum solution at 37 °C. Furthermore, this ATP-responsive MSN nanosystem was able to respond to both commercial ATP and ATP extracted from living cells. It was also suggested that the aptamers of other targets, such as ions, biological macromolecules and small molecules, cells and bacteria could be used in a similar manner to the ATP-aptamer for the fabrication of target-responsive MSN nanocarriers.

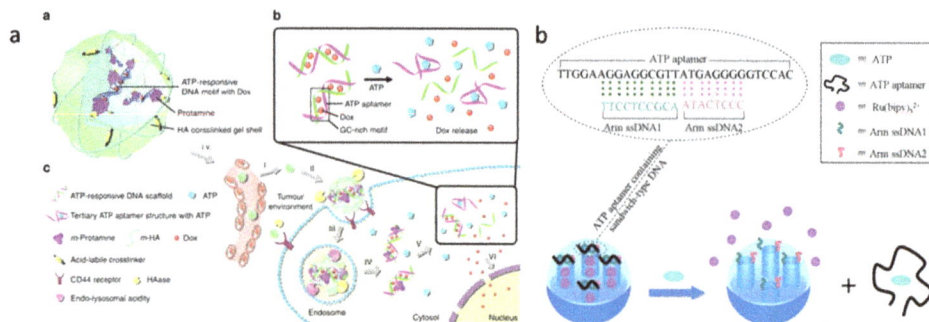

Figure 5.1. (a) Schematic illustration of the mechanism for targeted delivery and ATP-labile release of DOX from a DOX/nanogel; ATP-sensitive DNA motif with DOX, protamine and HA-crosslinked gel shell delivered into a tumor site through both passive and active targeting for cancer therapy. Reproduced with permission from [6]. Copyright 2014 Macmillan Publishers Ltd. (b) Schematic illustration of aptamer-functionalized ATP-triggered release of loaded cargos from MSN nanocarrier. Reproduced with permission from [2]. Copyright 2012 American Chemical Society.

5.3 Glucose-responsive

Functional MSN have been used in the design of DDS for the treatment of diabetes and cancers [7]. Glucose-triggered DDS could be applied for this purpose for diabetes. In type 2 diabetes, reduced or lost expression of glucose transporters has been reported and due to the high sugar concentrations and the inadequate production of insulin by the pancreas, glucose-responsive targeted drug delivery to particular cells could be required [8, 9]. Several cancers have also been associated with diabetes and with glucose metabolism [10] and the effects of glucose on cancer cells, as well as the overexpression of insulin and carbohydrate-based transporters in specific cancers [11], have been described. Biomolecules such as enzymes, proteins and functional insulin have been used for glucose-triggered gates with drug release capability. For example, glucose oxidase-gated MSN nanocarriers were developed on the basis of an enzyme-inhibition mechanism and are sensitive to very small concentrations of glucose. However, these approaches have many limitations in practical use. In recent studies, attempts have been made to design a glucose-triggered gate in MSN [7, 12]. Sinha *et al* [7] synthesized a dextran-gated multi-functional magnetic MSN with glucose-responsive gate-opening. The NP were functionalized with phenylboronic acid and folate (for targeting). In order to allow magnetic manipulation, imaging and tracking of the MSN, magnetic NP were loaded inside the mesopores. The low-cost and stable dextran molecule was used to prepare an adextran-based gating mechanism after the drug molecules were loaded into the pores. Then the surface of all the pores was closed via the binding of dextran with phenylboronic acid. Drug release was achieved by opening the dextran-gated pores, which was done by replacing the dextran with glucose molecules, which competitively bound to the phenylboronic acid (figure 5.2(a)). It was reported that polyethylene glycol (PEG)-like functionalization on the surface of the NP was

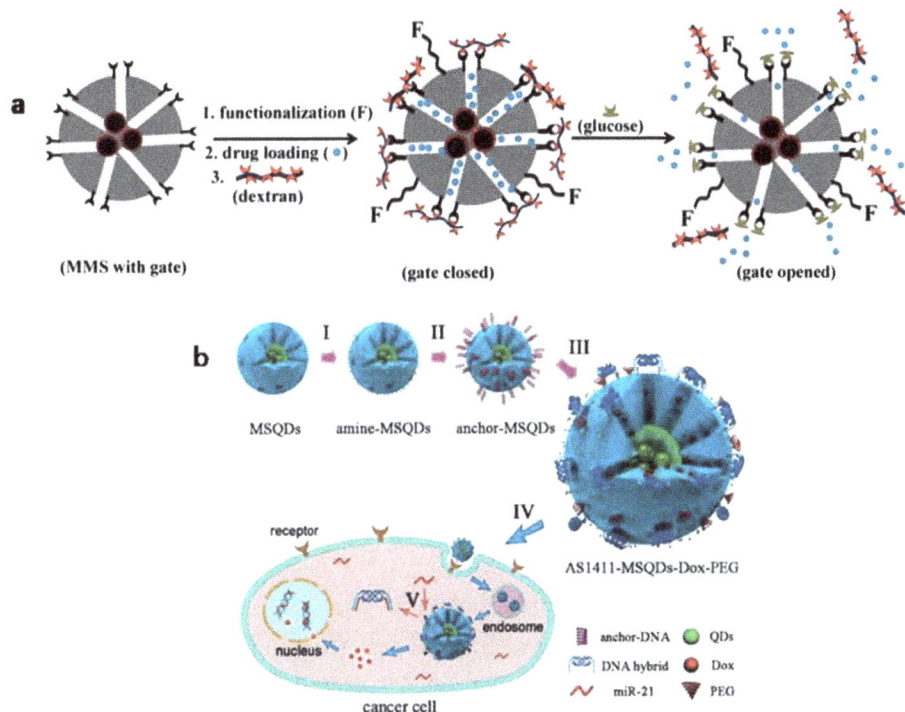

Figure 5.2. (a) Synthesis process for a dextran-gated multifunctional MSN nanocarrier, and glucose-sensitive gate-opening leading to drug release. Reproduced with permission from [7]. Copyright 2014 American Chemical Society. (b) Fabrication process for a DOX-loaded MSQD with functionalized amine moieties, hybrid capping with anchor DNA and PEGlation, followed by miRNA-triggered uncapping and cargo release after targeted recognition and endocytosis. Reproduced with permission from [13]. Copyright 2014 by John Wiley and Sons, Inc.

achieved through dextran-based gating, which provided specific targeting of cells with glucose receptors and reduced non-specific binding interactions. It was also shown that the nanocarrier was sensitive to bulk glucose concentration, providing good cellular delivery and release of type II diabetes drugs to target pancreatic beta cells, as well as glucose-dependent delivery of cancer drugs to cancer cells with folate receptors and subsequent cytotoxicity via folate functionalization of the nanocarrier.

5.4 DNA-responsive

DNA-responsive DDS have been developed to achieve effective release and specific targeting of tumor tissues. Maximum therapeutic efficiency with only minor side-effects can be obtained using aptamer-targeted delivery or microRNA (miRNA)-targeted drug release [13]. miRNA are short non-coding RNA molecules that regulate gene expression in diverse cellular processes, and are often overexpressed in various cancer tissues. Aptamer-targeted NP were developed by Min *et al* for the specific recognition of receptors (e.g. nucleotin) overexpressed on the tumor cell surface using MSN-coated quantum dots capped with a programmable DNA hybrid composed of an antisense oligonucleotideanti-miR-21 strand coupled with a DNA aptamer (i.e. AS1411). DOX was loaded within the nanocarrier and then the second structure of the aptamer in the DNA hybrid acted as a gatekeeper to keep the drug molecules encapsulated in the pores of the MSN. The specific uptake of nanocarriers by the target tumor cells was obtained through cell recognition by the AS1411 aptamer. Then exposure to miR-21 (an oncogenic miRNA overexpressed in diverse human cancers) led to the triggering of the anti-miR-21 by complementary base pairing, which unlocked the gate using a competing hybridization against the anchor DNA, thus allowing controlled drug release of miR-21 (figure 5.2(b)). Furthermore, the complementary base pairing between the anti-miR-21 and the miR-21 strands caused suppression of miR-21 expression, activating caspase-dependent apoptosis and subsequently killing the tumor cells. After aptamer-mediated recognition and endocytosis via delivery of DNA-hybrid-capped mesoporous silica-coated quantum dot (MSQD) nanocarriers into HeLa cells, the nanocarriers were unlocked by competitive hybridization of overexpressed endogenous miR-21 (as an exclusive key), with the DNA hybrid leading to the eradication of HeLa cells.

5.5 Reactive oxygen species-responsive

ROS such as oxygen free radicals and H_2O_2 are generated in most physiological processes. Normal cell functions are maintained using low concentrations of ROS that act through cellular signaling pathways, and homeostasis ensures that the ROS concentration is limited by antioxidant defenses. An enhanced concentration of ROS can lead to pathogenic processes in conditions such as aging, neurodegenerative diseases (e.g. Parkinson's, Alzheimer's and amyotrophic lateral sclerosis (ALS)), diabetes mellitus, carcinogenesis and atherosclerosis. Oxidative damage to biomolecules is deleterious (e.g. damage to DNA leading to mutations and different cancers) [14–17]. Recently, ROS have been used in biomolecule-responsive DDS for site-specific targeting and cargo delivery via nanocarriers that respond to the oxidative stress in different biological systems. ROS-responsiveness can be achieved through oxidative changes occurring in several materials, e.g. phase transitions, hydrolysis, a switch in solubility, or polymer degradation. Several systems have been developed to sense and repair the cellular damage [18] and obtain cell protection through stabilization of the mitochondrial membrane potential and reduction of intracellular ROS production [19]. Zhang *et al* [20] prepared biocompatible ROS-responsive NP from ROS-triggerable biocompatible β-cyclodextrin (β-CD) with low immunogenicity and *in vivo* safety. The nanocarrier was synthesized by the conjugation of 4-phenylboronic acid pinacol ester (PBAP) onto the hydroxyl groups of β-CD. Then the core–shell structure was designed using either self-assembly or a nano-precipitation method. It was possible to hydrolyze this nanocarrier into parent β-CD molecules via exposure to ROS conditions (figure 5.3(a)). The results indicated that DTX-loaded nanocarriers can lead to delay in drug dissolution and release, and can stimulate apoptotic activity against cancer cells and the inhibition of tumor growth. This nanosystem showed high ROS-sensitivity and good biocompatibility and it was proposed for the ROS-sensitive delivery of imaging agents and biomolecules.

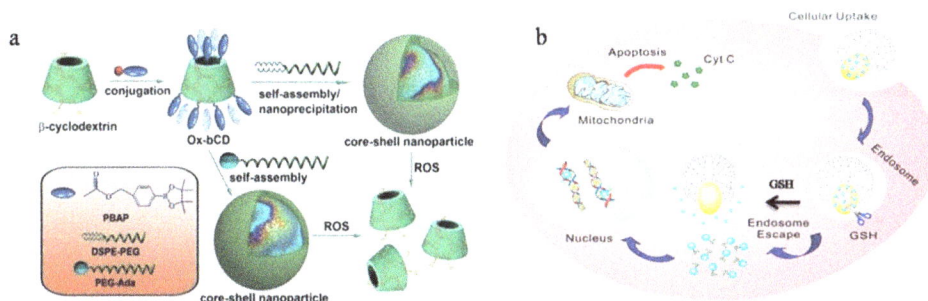

Figure 5.3. (a) Preparation process for a ROS-triggered β-CD to obtain (Ox–bCD)-based core–shell NP and subsequent cargo release via ROS-sensitive hydrolysis into β-CD molecules. Reproduced with permission from [20]. Copyright 2015 by John Wiley and Sons, Inc. (b) Endocytosis and GSH-triggered intracellular release of anticancer payloads from DEGNP nanocarriers. Reproduced with permission from [24]. Copyright 2013 American Chemical Society.

5.6 Glutathione-responsive

Glutathione is a nonprotein thiol in the cytoplasm of living cells [21]. Most therapeutic agents (e.g. anticancer drugs and anti-oxidants) and biotherapeutic agents (e.g. siRNA and protein and peptide drugs) perform effectively only inside the cell in intracellular compartments (e.g. nuclei and cytosol). Furthermore, the intracellular environment is rich in glutathione (GSH) tripeptides (2–10 mM) compared to extracellular environments (2–20 μM). Thus, glutathione can be used in stimuli-responsive DGDS when the cargo release is activated via the glutathione-sensitive destabilization of nanocarriers inside the cells [22]. Several glutathione-responsive nanovehicles have been developed, including hollow silica NP [23] and dendrimer-encapsulated gold NP [24], as well as capsules, polymersomes, nanogels, macromolecular drug conjugates and micelles [22]. In one study, glutathione-sensitive 'on–off' release was achieved using a biocompatible nanovehicle composed of dendrimer-encapsulated gold NP (DEGNP) that allowed the release of thiolated anticancer drugs (e.g. cisplatin and DOX). The thiol-containing drugs such as captopril and 6-mercaptopurine (sensitive to glutathione) were loaded into the DEGNP nanocarriers. These nanocarriers can be loaded with the thiolated drugs through the formation of Au–S bonds. In the presence of thiol-reducing agents such as glutathione and dithiothreitol, the loaded drug molecules were released (figure 5.3(b)). The results showed reduced cytotoxicity for drug-loaded nanocarriers compared to the free anticancer drug, but this could be increased by raising the glutathione level inside the cells. It was suggested that using PEGylated dendrimers as templates for gold NP could enhance the drug-loading capacity, prolong the *in vivo* blood circulation time and allow them to be used as contrast agents for computed tomography imaging [24]. In another study, thiol-responsive amphiphilic gemini micelles composed of hydrophilic PEG blocks and hydrophobic polylactide (PLA) blocks joined by a cystine disulfide spacer were prepared and used for intracellular DOX drug delivery. In the intracellular milieu, which is rich in glutathione, the gemini micelles destabilized into monomeric micelles upon cleaving the cystine linkages. This resulted in alterations in the size distribution of the micelles and the formation of aggregates and, more importantly, to increased intracellular release of encapsulated DOX compared to the minimal release under non-reductive conditions. Thus, non-toxic $(PEG)_2–Cys–(PLA)_2$ nanocarriers showed enhanced anticancer efficacy and controlled drug delivery [25].

5.7 Receptor-responsive

Receptor-mediated DDS have attracted interest recently because of their particular advantages, especially for the treatment of neurological disorders such as Alzheimer's disease, stroke and multiple sclerosis, where they can target the brain by crossing the physical obstacle of the blood brain barrier (BBB). The endogenous receptor-mediated transport of biologics modified with appropriate targeting ligands via vesicular trafficking is a promising strategy for overcoming the BBB [26]. Receptor-mediated DDS with noncovalent targeted gold NP for simultaneous brain tumor therapy and noninvasive imaging have been developed. Dixit et al [27] prepared transferring-peptide (Tf_{pep}) receptor-coated PEGylated theranostic gold NP (AuNP) loaded with the photodynamic prodrug Pc4 for photosensitizer delivery to brain cancer cell lines and orthotopic brain tumors to study the efficiency of the therapeutic DDS in crossing the BBB endothelium. Targeted delivery of Tf_{pep}–AuNP clusters to human glioma cancer lines (LN229 and U87) overexpressing the transferrin receptor (TfR) led to strong enhancement of the AuNP specificity and cellular uptake because of the targeting capabilities of the Tf_{pep} ligand compared to untargeted particles. The Tf_{pep}-conjugated Pc4-loaded AuNP delivered the Pc4 effectively within the vesicles. These nanocarriers were internalized in Tf-expressing endosomes and accumulated successfully within the mitochondria of glioma cancer cells with insignificant non-specific accumulation. This led to efficient PDT killing of the cancer cells by Pc4.

5.8 Cytoplasm-responsive

Cytoplasm-responsive delivery systems have been studied recently. Okada et al reviewed these systems for the delivery of siRNA employing cell-penetrating peptide nanomicelles [28]. Using small interfering RNA (siRNA) to silence gene expression through sequence-specific posttranscriptional gene silencing (PTGS) is a popular strategy for the treatment of various gene-related diseases, such as age-related macular degeneration (AMD), respiratory syncytial virus (RSV) infection, diabetic macular edema (DME), other viral infections, metastatic cancers, blood diseases, asthma, hypercholesterolemia, HIV, neurological disorders, cardiovascular diseases, metabolic disorders, etc [29–31]. Tanaka et al [32] produced a gene carrier for the treatment of cancer through the systemic injection and delivery of siRNA. They synthesized a cytoplasm-responsive nanocarrier composed of methoxypolyethyleneglycol–polycaprolactone (MPEG–PCL) diblock copolymers conjugated with a cytoplasm-responsive cell-penetrating peptide (CPP), $CH_2R_4H_2C$. This amphiphilic cationic nanocarrier was able to deliver and release siRNA effectively into S-180 cells in vivo and in vitro in the presence of fetal bovine serum (FBS). The intravenous injection of the $MPEG–PCL–CH_2R_4H_2C$ complexes avoided enzymatic degradation (i.e. it stabilized siRNA against nucleases (RNase)) by the disulfide linkage with artificial CPP, as well as rigid compaction of siRNA by ionic interactions. The nanocarrier showed good stability in blood circulation after intravenous injection. This strategy also avoided uptake of the nanocarriers by the reticuloendotherial system (RES) due to PEGylation of the nanomicelles that could be localized in tumors by the EPR effect. This approach allowed efficient cellular uptake and release of siRNA inside the tumor (after cleavage of the disulfide linkages in the reducing cytosol), while the siRNA (siVEGF) was able to prevent vascular endothelial growth factor (VEGF) secretion from S-180 cells (i.e. siRNA silencing via $CH_2R_4H_2C/$ siVEGF complexes) compared to naked siVEGF. The antitumor efficacy and tumor growth inhibition in S-180 tumor-bearing mice were enhanced. Hence, the systemic delivery of siRNA into tumor cells by the cytoplasm-triggered MPEG–PCL–$CH_2R_4H_2C$ nanocarriers was demonstrated, with the advantages of non-cytotoxic siRNA nanocarriers, higher uptake and early endosomal escape. It was suggested that this novel nanocarrier could be used for the clinical delivery of genes.

References

[1] Miyata T, Uragami T and Nakamae K 2002 Biomolecule-sensitive hydrogels *Adv. Drug Deliv. Rev.* **54** 79–98

[2] He X, Zhao Y, He D, Wang K, Xu F and Tang J 2012 ATP-responsive controlled release system using aptamer-functionalized mesoporous silica nanoparticles *Langmuir* **28** 12909–15

[3] Mo R, Jiang T and Gu Z 2014 Enhanced anticancer efficacy by ATP-mediated liposomal drug delivery *Angew. Chem.* **126** 5925–30

[4] Mo R, Jiang T, Sun W and Gu Z 2015 ATP-responsive DNA–graphene hybrid nano-aggregates for anticancer drug delivery *Biomaterials* **50** 67–74

[5] Sun W, Mo R, Jiang T and Gu Z 2014 *ATP-Responsive Hybrid DNA/Graphene Nanoassemblies for Anticancer Drug Delivery Biomaterials* **50** 67–74

[6] Mo R, Jiang T, DiSanto R, Tai W and Gu Z 2014 ATP-triggered anticancer drug delivery *Nat. Commun.* **2014** 5

[7] Sinha A, Chakraborty A and Jana N R 2014 Dextran-gated, multifunctional mesoporous nanoparticle for glucose-responsive and targeted drug delivery *ACS Appl. Mater. Interf.* **6** 22183–91

[8] Thorens B, Weir G C, Leahy J L, Lodish H F and Bonner-Weir S 1990 Reduced expression of the liver/beta-cell glucose transporter isoform in glucose-insensitive pancreatic beta cells of diabetic rats *Proc. Natl Acad. Sci.* **87** 6492–6

[9] Thorens B, Wu Y, Leahy J L and Weir G C 1992 The loss of GLUT2 expression by glucose-unresponsive beta cells of db/db mice is reversible and is induced by the diabetic environment *Journal of Clinical Investigation* **90** 77

[10] Gristina V, Cupri M G, Torchio M, Mezzogori C, Cacciabue L and Danova M 2015 Diabetes and cancer: a critical appraisal of the pathogenetic and therapeutic links (review) *Biomed. Rep.* **3** 131–6

[11] Xu C-X, Zhu H-H and Zhu Y-M 2014 Diabetes and cancer: associations, mechanisms and implications for medical practice *World J. Diabetes* **5** 372

[12] Wu S, Huang X and Du X 2013 Glucose-and pH-responsive controlled release of cargo from protein-gated carbohydrate-functionalized mesoporous silica nanocontainers *Angew. Chem.* **125** 5690–4

[13] Zhang P *et al* 2014 DNA-hybrid-gated multifunctional mesoporous silica nanocarriers for dual-targeted and microRNA-responsive controlled drug delivery *Angew. Chem. Int. Ed.* **53** 2371–5

[14] Yu B P 1994 Cellular defenses against damage from reactive oxygen species *Physiol. Rev.* **74** 139–62

[15] Waris G and Ahsan H 2006 Reactive oxygen species: role in the development of cancer and various chronic conditions *J. Carcin.* **5** 14

[16] Ray P D, Huang B-W and Tsuji Y 2012 Reactive oxygen species (ROS) homeostasis and redox regulation in cellular signaling *Cell. Signal.* **24** 981–90

[17] Apel K and Hirt H 2004 Reactive oxygen species: metabolism, oxidative stress and signal transduction *Annu. Rev. Plant Biol.* **55** 373–99

[18] Lee S H, Gupta M K, Bang J B, Bae H and Sung H J 2013 Current progress in reactive oxygen species (ROS)-responsive materials for biomedical applications *Adv. Healthc. Mater.* **2** 908–15

[19] Yin J-J *et al* 2009 The scavenging of reactive oxygen species and the potential for cell protection by functionalized fullerene materials *Biomaterials* **30** 611–21

[20] Zhang D *et al* 2015 Biocompatible reactive oxygen species (ROS)-responsive nanoparticles as superior drug delivery vehicles *Adv. Healthc. Mater.* **4** 69–76

[21] Ock K *et al* 2012 Real-time monitoring of glutathione-triggered thiopurine anticancer drug release in live cells investigated by surface-enhanced Raman scattering *Anal. Chem.* **84** 2172–8

[22] Cheng R, Feng F, Meng F, Deng C, Feijen J and Zhong Z 2011 Glutathione-responsive nano-vehicles as a promising platform for targeted intracellular drug and gene delivery *J. Controlled Release* **152** 2–12

[23] Wang D *et al* 2014 Fabrication of single-hole glutathione-responsive degradable hollow silica nanoparticles for drug delivery *ACS Appl. Mater. Interf.* **6** 12600–8

[24] Wang X *et al* 2013 Glutathione-triggered 'off–on' release of anticancer drugs from dendrimer-encapsulated gold nanoparticles *J. Am. Chem. Soc.* **135** 9805–10

[25] Kim H-C *et al* 2015 Thiol-responsive gemini poly (ethylene glycol)-poly (lactide) with a cystine disulfide spacer as an intracellular drug delivery nanocarrier *Colloid. Surface. B* **127** 206–12

[26] Lajoie J M and Shusta E V 2015 Targeting receptor-mediated transport for delivery of biologics across the blood–brain barrier *Annu. Rev. Pharmacol. Toxicol.* **55** 613–31

[27] Dixit S, Novak T, Miller K, Zhu Y, Kenney M E and Broome A-M 2015 Transferrin receptor-targeted theranostic gold nanoparticles for photosensitizer delivery in brain tumors *Nanoscale* **7** 1782–90

[28] Okada H, Ogawa T, Tanaka K, Kanazawa T and Takashima Y 2014 Cytoplasm-responsive delivery systems for siRNA using cell-penetrating peptide nanomicelles *J. Drug Deliv. Sci. Technol.* **24** 3–11

[29] Zhou J, Shum K-T, Burnett J C and Rossi J J 2013 Nanoparticle-based delivery of RNAi therapeutics: progress and challenges *Pharmaceuticals* **6** 85–107

[30] Kole R, Krainer A R and Altman S 2012 RNA therapeutics: beyond RNA interference and antisense oligonucleotides *Nat. Rev. Drug Discovery* **11** 125–40

[31] Burnett J C and Rossi J J 2012 RNA-based therapeutics: current progress and future prospects *Chem. Biol.* **19** 60–71

[32] Tanaka K *et al* 2013 Cytoplasm-responsive nanocarriers conjugated with a functional cell-penetrating peptide for systemic siRNA delivery *Int. J. Pharmaceut.* **455** 40–7

Mahdi Karimi, Parham Sahandi Zangabad, Amir Ghasemi and Michael R Hamblin

Chapter 6

Dual/multi-stimuli-sensitive nanocarriers

6.1 Introduction

Combinations of two or more of the various stimuli discussed in the preceding sections can be used to further enhance the versatility and specificity of triggerable DGDS [1, 2]. Such dual/multi-functional stimuli-responsive nanomaterials can be even smarter and also possess higher loading efficiency and more prolonged sustained release times than their mono-responsive counterparts. In addition, these nanomaterials may be capable of sensing only slight alterations in their environment (e.g. pH or reductive/oxidative potential variations), and this may even extend to external stimuli, such as temperature, light or magnetic field [3–7].

Dual- and multi-responsive smart nanocarriers could solve some of the general challenges of specific drug-delivery, including prolonging stability, better cellular internalization and uptake, and intracellular release of the drug molecules. Various combinations of different types of stimulation can be combined, including external and internal stimulation, and they can occur at separate different times or they can take place simultaneously [8].

Single stimulus-responsive DDS can have several drawbacks. Co-delivery of genes and drugs with synergistic effects has been explored for the treatment of several difficult diseases, such as cancers. However, the time gap between the onset of the function of the delivered gene therapy (i.e. transcription and translation of genetic materials, or prevention of protein expression), which tends to be longer acting, and the time point of action for therapeutics (which tend to be shorter acting) has to be considered in the design of combination therapies that co-deliver genes and drugs [8]. In this respect, dual/multi-responsive systems are of great importance in achieving controlled and sequential release of genes and drugs. By applying combinations of various external and internal stimuli, the triggered internalization and release of genes and therapeutics in intracellular compartments and the initiation of their function can be accomplished in a time-regulated manner.

In recent years, miscellaneous dual-responsive DGDS have been developed. In this section, we review some of the latest dual- and multi-responsive nanocarriers. Table 6.1 shows the most recent DGDS that use combinations of internal stimuli with each other or with external stimuli for the triggered delivery and release of cargos.

6.2 Dual stimuli-based delivery systems

6.2.1 pH/temperature-responsive

DDS with a simultaneous response to both pH and temperature stimulation have been studied [16–20]. Various types of pH/temperature-responsive nanocarriers have been designed, including temperature/pH-responsive polymeric NP [21], pH–thermo-sensitive microcontainers [22] and pH–thermo-sensitive core–shell NP [23]. In a study by Rodkate et al [24], a thermo-sensitive polymer (i.e. poly (NIPAAm) and a pH-triggered polymer (i.e. poly(DEAEMA)) were conjugated together with carboxymethylchitosan (CMC) by radical polymerization, leading to a hydrogel structure. This structure was used as an ON/OFF switch. Thus it was indicated that the hydrogels underwent water swelling at either basic pH conditions (pH 11.0) or below the lower critical solution temperature LCST transition (~32 °C) at aqueous solution, both of which can be regulated to mediate entrapped drug molecules. Salehi et al [16] reported on the pH/thermo-responsiveness of a DOX/MTX-loaded MSN-based nanocarrier, which was fabricated by grafting an ionic liquid thermopolymer onto the MSN. This DOX/MTX-loaded nanocarrier was sensitive to simultaneous acidic milieu and high temperature, and showed high cytotoxicity against A549 and MCF7 cancer cell lines, in contrast to the free nanocarrier. Mu et al designed a dual thermo–pH-triggered nanocapsule that used tert-butylacrylate (t-BA) (located at the inner walls of the nanocapsule) and was coated with NIPAm and N,N-methylenebisacylamide (MBA). A PNIPAm crosslinked shell was the thermo-sensitive compartment of the nanocapsule and poly-t-BA brushes formed the pH-sensitive compartment. The ester moieties of t-BA were then hydrolyzed into carboxyl groups. The nanocapsule was able to carry drugs possessing hydroxyl or amine groups [25]. In another study, in an effort to enhance the half-life and bioavailability of fibroblast growth factor2 (FGF2) (a protein coding gene), a pH/temperature-responsive poly (NIPAAM-co-PAA-co-BA) microsphere was synthesized and the delivery of therapeutic proteins to ischemic skeletal muscle was investigated. Sustained, diffusion-controlled release of FGF2 within ischemic tissues was obtained at the simulated pH of ischemic muscle (pH 5.2–7.2), while under physiological conditions (pH 7.4) the dissociation and consequent rapid clearance of microspheres occurred [26]. In another study, a copolymer of urethane, poly-amino-urea-urethane (PAUU), was used for sustained delivery of proteins (e.g. FITC-BSA). This nanocarrier underwent a sol-transition at environmental temperature and in acidic conditions, while at physiological temperature and pH it was converted to the gel form [27].

Table 6.1. Several combinations of internal stimuli integrated either with each other or with external stimuli for use in DGDS.

Stimuli	Nanocarrier	Synthesis method	Cargo	Results and outcome	Reference
Photo–redox	Dual-responsive reversibly crosslinked micelles for anticancer targeted drug delivery in which the coumarin and disulfide bonds are introduced into a linear–dendritic copolymer (telodendrimer) through peptide chemistry	Formation of micelles via tuning of the telodendrimer architectures to self-assemble by introducing amphiphilic CA at the adjacent α-amino and conjugating coumarin on the ε-amino on polylysine via a disulfide bond-containing spacer	Anticancer PTX	• Crosslinking via controllable coumarin photodimerization within the nanocarrier • De-crosslinking of the nanocarriers via UV irradiation with various wavelengths in addition to simultaneous cleavage of crosslinked nanocarriers by redox-triggered built-in disulfide crosslinkage in the tumor microenvironment, leading to on-demand drug release • Good drug-loading capacity and stability • Similar anticancer efficacy • Reduced toxic side-effects due to stable drug encapsulation during incubation with cells • Good tumor accumulation and long-term residency in tumor of the delivered drugs by the crosslinked micelles due to their increased mechanical stability	[9]
Thermo–redox	Poly(N-isopropylacrylamide) (PNIPAM)-co-2-hydroxyethyl acrylate) hydrogels crosslinked via diselenides	Crosslinking copolymers of N-isopropylacrylamide (NIPAM) and 2-hydroxyethyl acrylate (HEA) with a diselenide-bearing	Model drug salicylic acid (SA)	• Swelling–shrinkage behavior of hydrogels triggered via external temperature due to hydrophobic–hydrophilic transition of PNIPAM segments; elevated temperature resulted in the	[10]

(Continued.)

Table 6.1. (Continued.)

Stimuli	Nanocarrier	Synthesis method	Cargo	Results and outcome	Reference
		crosslinking agent via free radical copolymerization of NIPAM and HEA		shrinkage of the hydrogels due to lower critical solution temperature (LCST) phase transition of PNIPAM • Oxidation-triggered gel-to-sol transition of the hydrogels; oxidizing agent (H_2O_2) triggered the dissolution of the diselenide crosslinked hydrogels into polymer solutions • Temperature-induced slow sustained drug release and oxidation-induced rapid burst release	
pH–redox	Biodegradable monodisperse PEGylated poly [methacrylic acid-co-poly(ethylene glycol) methyl ether methacrylate (PEGMA)-co-N,N-bis (acryloyl)cystamine], (PMPB), spherical nanohydrogels (200 nm)	Facile and clean one-step distillation precipitation polymerization	Doxorubicin hydrochloride	• Adequate drug loading capacity and high drug encapsulation efficiency (more than 96%) • Negligible premature drug release during blood circulation • Quick drug release at reduced pH conditions and through exposure to reductive reagent glutathione (GSH) and subsequent cleavage of the disulfide crosslinking bonds • Cumulative release of more than 85% in 30 h • PEGMA reduced cytotoxicity leading to non-toxicity of nanohydrogels against HepG2 cells at concentrations equal to or lower than 10 μg mL^{-1}	[11]

Table 6.1. (Continued.)

Stimuli	Nanocarrier	Synthesis method	Cargo	Results and outcome	Reference
pH–electrical	Bacterial cellulose nanofiber/sodium alginate (SA) hybrid hydrogels (nf–BC) with pH and electric-field-responsive swelling	The BC slurry and sodium alginate solution were homogeneously mixed at various weight ratios and then crosslinked by an aqueous solution (CaCl$_2$); finally, a freeze-drying process was applied	Model drug (Ibuprofen)	• Increasing pH from 1.5 to 11.8 enhanced the swelling ratio • Changing the electric field from 0 to 0.5 V increased the swelling ration of the hydrogels • Deprotonation or protonation of calcium alginate in the hydrogels under different pH conditions can control the release rate of Ibuprofen; rapid release in neutral or alkaline conditions and lower release rate in acidic conditions • Electric field enhanced the drug release	[12]
Photothermal–pH	β-Cyclodextrin/paclitaxel complexed fluorescent carbon-based NP (partially carbonized fluorescence HA (HA–FCN))	Boronic acid conjugated HA-FCN (HA-FCN–BA) (leading to formation of boronate ester diol groups of β-cyclodextrin (CD) [HA-FCN–CD]) was synthesized from 3-amino phenyl boronic acid and carboxyl group activated HA-FCN; here, the dehydration of HA (and carbonizing HA) led to formation of FCN, and HA-FCN was constructed via a lower degree of carbonization	Anticancer PTX	• Nanocarrier used for targeted cancer treatment and bioimaging • Optical absorption similar to that of NIR-irradiation triggered carbonized materials • Acidic milieu triggered the drug release and mild photothermal heat induced burst release of PTX	[13]

(Continued.)

Table 6.1. (Continued.)

Stimuli	Nanocarrier	Synthesis method	Cargo	Results and outcome	Reference
Light–pH	pH-responsive HA spherical NP (PHAN) (sub-100 nm size and negative surface charge) for spatiotemporally controlled cytosolic drug release	PHAN were synthesized via the self-assembly of a photosensitizer (chlorin e6) and a pH-sensitive moiety (poly-(diisopropylaminoethyl) aspartamide (PDIPASP)) conjugated to HA	DOX	• pH changes triggered the disassembly of the NP due to the protonation of PHAN, leading to DOX release • Low-intensity laser irradiation triggered the free PS to produce reactive singlet oxygen, leading to the release of DOX within the cytosol of cancer cells • CD44-receptor-mediated endocytosis, endosomal escape capability and efficient drug targeting • Enhanced anticancer efficacy of dual-triggered DOX@PHAN over free DOX and PHAN	[14]
pH–biomolecule	Bioresponsive polymer MSN as an 'AND' logic gate with immobilized polycaprolactone (esterase degradable) in the core and polyacrylic acid (PAA) as pH-responsive agent outside the MSN	The classical cetyltrimethylam-monium bromide (CTAB)-templated, base-catalyzed sol–gel method for synthesis of MSN	DOX	• Payload released only in the presence of both pH (acidic conditions) and esterase found in tumors (AND logic gate); indicative of unique physiological milieu of solid tumors • pH stimulus cleaved the hydrogen bonds, leading to surface charge switching • Enzymatic stimulus induced hydrolysis of ester bonds • Cellular uptake of the intact nanocarrier followed by the intracellular delivery of DOX into the nucleus • AND logic gate delivery system was highly efficient (more than eightfold) against neuroblastoma cells compared to normal fibroblasts	[15]

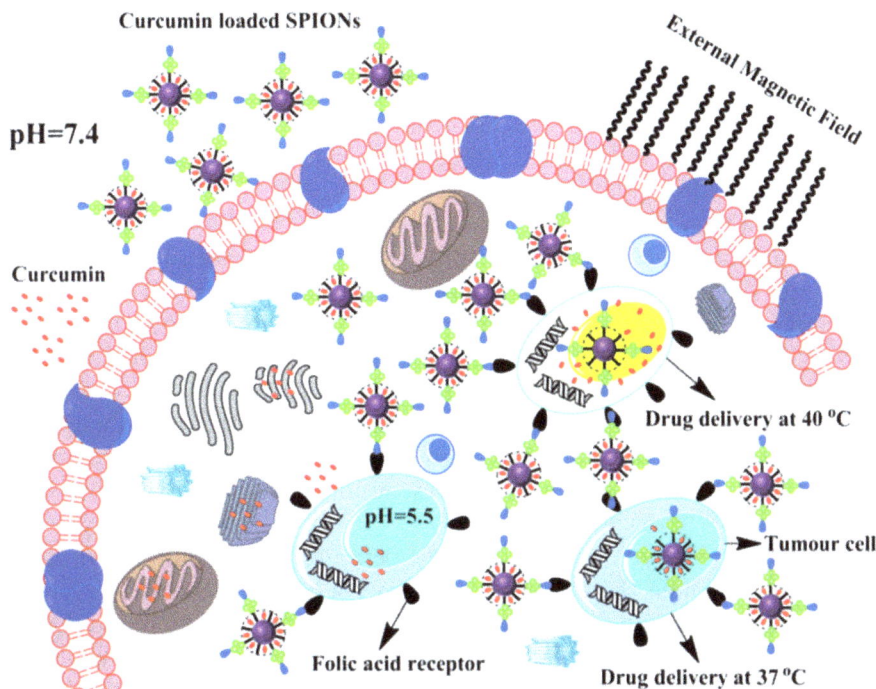

Figure 6.1. Dual pH- and temperature-sensitive PNIMAAm–co–PGA-coated curcumin-loaded nanocarrier equipped with SPION for enhanced targeted delivery: schematic illustration of cellular uptake of the nanocarrier by MCF67 cancer cells and intracellular release at pH 5.5 and LCST 40 °C followed by localized hyperthermia at the higher temperatures. Reproduced with permission from [29]. Copyright 2015 American Chemical Society.

Sundaresan *et al* [28] showed that using pH reduction and temperature increase for poly(N-isopropylacrylamide-acrylamide-chitosan) (PAC)-coated magnetic NP, the maximum drug release rate was achieved at pH 6.0 and 40 °C, respectively. In another study, Patra *et al* [29] synthesized water-soluble superparamagnetic iron oxide NP (SPION) (15–20 nm) loaded with the hydrophobic anticancer drug, curcumin. This nanocarrier was coated with dual pH–temperature-responsive poly-N isopropyl acrylamide–co–poly-glutamic acid (PNIMAAm–co–PGA). The polymer-coated nanocarrier showed high drug-loading capacity (89%) and encapsulation efficiency (98%). The nanocarriers were taken up efficiently by MCF67 cancer cells and the drug molecules were released rapidly at pH 5.5 and temperature 40 °C (LCST transition) (figure 6.1). Through magnetic field irradiation, the SPION temperature was increased to 45 °C, leading to rapid and efficient localized hyperthermia.

An injectable dual-responsive hydrogel was fabricated for applications in tissue engineering using a combination of human-like collagen–chitosan (HLC–CS) and beta-sodium glycerophosphate (HLC–CS/beta-GP), which was triggered at physiological pH and a temperature of around 36–38 °C. Temperature-responsiveness was obtained through the hydrophobic–hydrophilic interactions between the –CONH,

–OH and –COOH groups and the alkenes, cycloparaffins and hydrocarbons of the hydrophobic moieties, while pH-sensitivity was obtained through the electrostatic interactions between the $-NRH_2^+$ and $-OPO_3^{2-}$ moieties. The presence of HLC reduced the release time noticeably, while it also enhanced the swelling ratio and mechanical strength, as well as the porous structure and porosity of the hydrogel [30].

GO has been used in targeted delivery systems because it has a large surface area and can be functionalized with targeting ligands. GO nanosheets have been utilized in the design of dual- and multi-stimuli-triggered nanocarriers [31–34]. A hybrid hydrogel of GO nanosheets and PNIPAAm was designed that possessed reversible sensitivity to both temperature and pH [35]. Shrinkage was obtained via heating from 20 °C to 50 °C and the nanovehicles returned to their initial volume with a rapid (10 s) cooling process. The pH-triggered behavior was attributed to the existence of residual carboxyl moieties in the hydrogel which formed hydrogen bonds that were ionized at high pH values, resulting in swelling of the structure (figure 6.2).

Shim *et al* [36, 37] synthesized a dual temperature/pH-responsive hydrogel that was composed of three thermosensitive polymers—poly(ε-caprolacton-co-lactide), poly ethylene glycol and poly(ε-caprolacton-co-lactide) (poly(CL-co-LA)–PEG–poly-(CL-co-LA))—and conjugated to acidic sulfamethazine oligomers (SMO). The degradation rate of the pH- and thermo-triggered block copolymer was significantly reduced compared to the control block copolymer as a result of the buffering effect of the sulfamethazine oligomers–poly(ε-caprolactocton-co-lactide)–poly ethylene glycol–poly(ε-caprolacton-co-lactide) (SMO–PCLA–PEG–PCLA–SMO) sulfonamide moieties on the acidic monomer. This eliminated the critical drawbacks of thermo-sensitive block copolymers, including premature gelation and rapid degradation. In addition, a sustained release of PTX for 2 weeks after a single injection was provided (figure 6.3).

A dual-sensitive hydrogel was constructed via binding of PEG to 4-poly(beta-amino ester) (PAE) to form a hydrogel *in vivo* after injection (figure 6.4(a)) [38]. Modification of the PAE structure resulted in the generation of a new hydrogel, oligo (beta-amino ester-urethane) (OAEU), which was capable of sustained DOX release over 10 days in rats (figure 6.4(b)) [39].

6.2.2 Light/pH-responsive

Dual light/pH-responsive systems have also captured the interest of researchers and, after temperature–pH systems, they are one of the most common combinations [40]. In one study, dual-stimulus-responsive MSN functionalized with propyl-carbamic acid 4,5-dimethoxy-2-nitrobenzyl ester (or nitroveratryl carbamate-protected amino-propyl (NVCAP)-functionalized MSN) resulted in the final structure, mercaptopropyl (MP)–MSN, which responded to both pH and light [41]. In another attempt, Feng *et al* [42] used an azobenzene copolymer as a photosensitive material; this was combined with acrylic acid and then a self-assembly process followed. This resulted in a dual photo/pH-responsive nanoparticle.

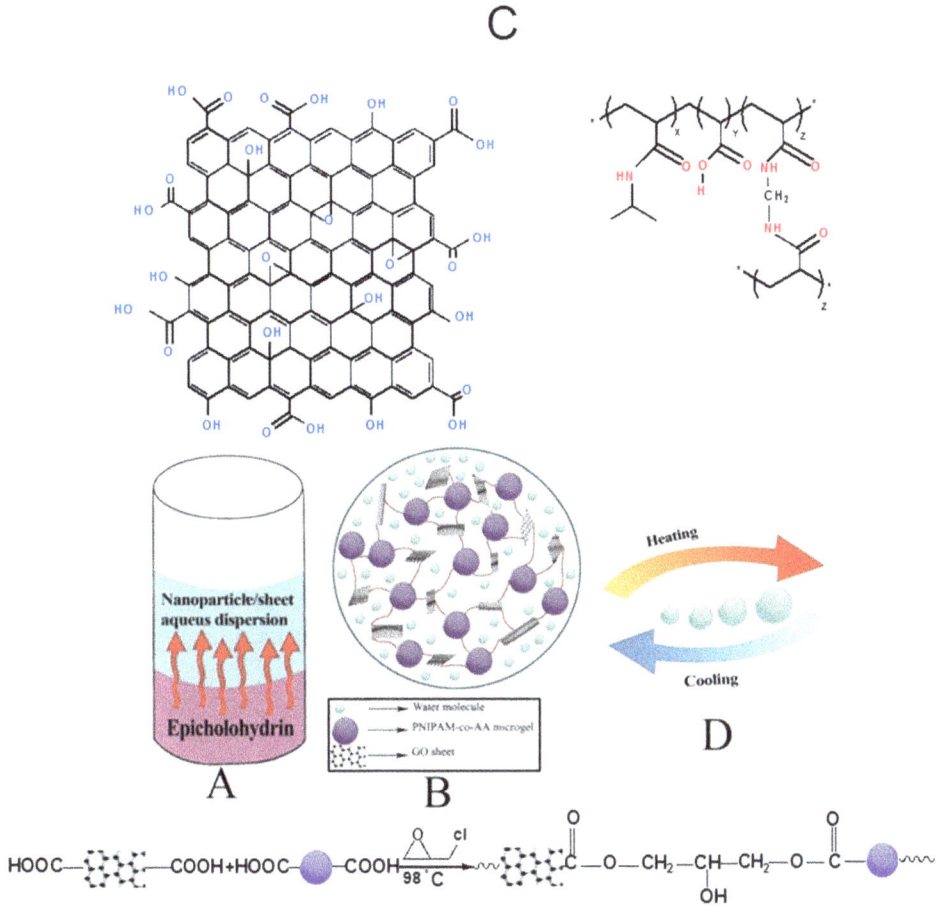

Figure 6.2. (a) Synthesis procedure for the hydrogel; (b), (c) the structure and formula; and (d) the response to temperature changes.

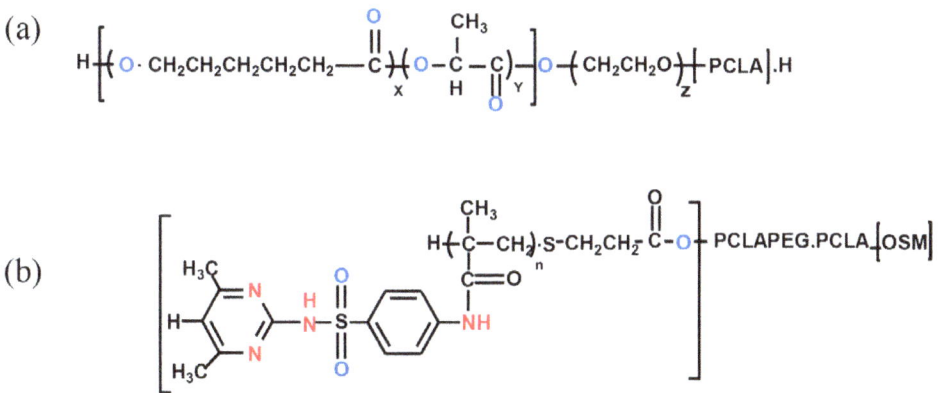

Figure 6.3. Structure of (a) PCLA–PEG–PCLA polymers and (b) SMO–PCLA–PEG–PCLA–SMO polymers.

(a)

(b)

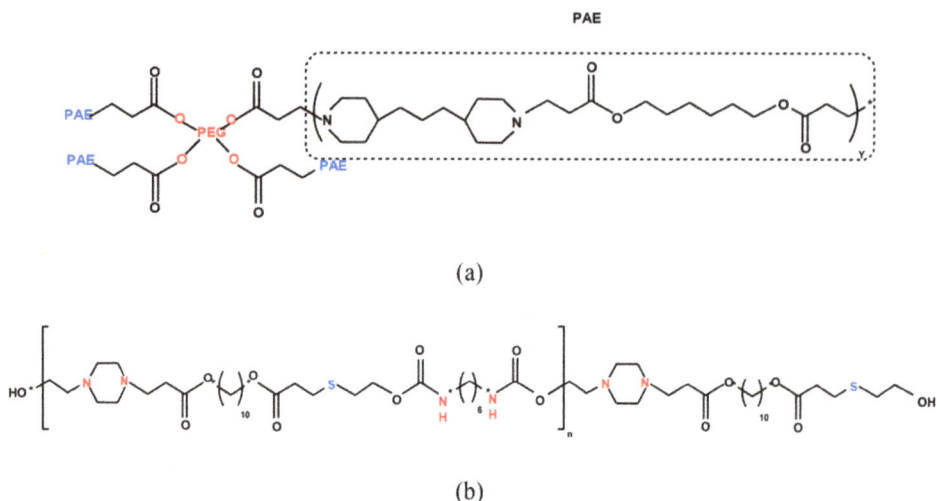

Figure 6.4. Structure of the three basic blocks of biodegradable double pH/temperature-sensitive hydrogel: (a) PEG–4PAE and (b) oligo OAEU.

Dual-responsive NP can also be used for simultaneous imaging and therapy. Yu *et al* [43] fabricated a dual-functional pH-sensitive and NIR-emissive dextran NP modified with pH-susceptible pendant acetals. The nanocarriers were degraded in mildly acidic conditions by hydrolysis of the dextran and simultaneously the NIR fluorescence probe indicated the DOX release process via a Forster resonance energy transfer (FRET) mechanism between DOX (as the excited donor) and BTTPF (the electron acceptor).

6.2.3 Magnetic/pH-responsive

Another class of dual-responsive systems uses magnetic field targeting and pH triggering for drug delivery. A hyperthermia-based temperature-responsive/ magnetic-targeted dual-responsive complex carrier was designed by combining pH-sensitive chitosan NP with thermo-sensitive amine-terminated poly-N-isopropylacrylamide. Gadolinium doped nickel ferrite (the magnetic-guidance agent) was encapsulated in the carrier complex and fluorescein was used as a fluorescent marker, targeted with tagged folic acid (FA). Curcumin was loaded in the nanocarrier with a drug loading efficiency of 86%. In addition, incubation with four cancer cell lines resulted in apoptosis [44].

A composite of SiO_2–Fe_3O_4 was encapsulated by mPEGpoly(l-asparagine) to construct a three-layered core–shell microsphere for the delivery of DOX. The imidazole moiety of asparagine acted as the pH-responsive moiety, inducing release of DOX at a pH of 5.5, whereas mPEG enhanced the blood circulation lifetime [45]. Kang *et al* synthesized a monodisperse poly(acrylic acid)-modified Fe_3O_4 ($PAA@Fe_3O_4$) hybrid microsphere that was sensitive to pH and magnetic field. Here, PAA was encapsulated in the Fe_3O_4 hollow spheres and then DOX was loaded into polyacrylic acid–Fe_3O_4 microspheres. The results showed the strong dependence of the hybrid microspheres and the subsequent DOX release on the pH value due to the presence of the PAA polymer. Under magnetic guidance, greater cellular uptake of the DOX-loaded microspheres was shown by HeLa cells with an acidic environment compared to normal cells, and this led to enhanced cytotoxicity [46]. Yang *et al* used β-thiopropionate-poly(ethylene glycol) as the pH-sensitive gatekeeper part of the $Fe_3O_4@mSiO_2$ core–shell nanocomposite nanocarrier, which was then loaded with DOX. The nanocarriers were targeted towards the cancer sites via magnetic guidance. Thereafter, selective pH-triggered release was obtained through the hydrolysis of the β-thiopropionate linkage upon exposure to mildly acidic conditions. This system showed good avoidance of premature drug release. In addition, the PEG-modified nanocarriers of small size (about 65 nm) were taken up efficiently by HeLa cells, resulting in an effective cancer therapy [19].

A magnetic-targeted, pH-triggered emulsion droplet-based delivery system was studied by Mu *et al* [47]. Cationic oligo–chitosan and anionic sodium alginate were used to construct a drug carrier via layer-by-layer self-assembly of polyelectrolytes onto drug-loaded emulsion droplets induced by electrostatic interaction, with hybrid emulsion droplets encompassing superparamagnetic ferroferic oxide NP and drug dipyridamole (DIP) molecules as their cores. Thereafter, the polyelectrolyte solution was removed to form magnetic–DIP–oleic acid NP. High drug encapsulation efficiency and a cumulative release ratio of almost 100% was obtained from the oligochitosan/sodium alginate multilayer-encapsulated magnetic hybrid emulsion droplets at pH 1.8, compared to a cumulative release rate of 3.3% at pH 7.4.

6.2.4 pH/redox-responsive

The acidic pH and high concentration of ROS in cancer cells indicate that oxidative environments are a desirable target for the delivery of anticancer drugs [48–51]. A thiol-modified self-assembled carboxymethyl chitosan crosslinked nanoparticle was synthesized with an average size of 160 nm which showed a triggered release of methotrexate drug byredox and acid stimulations [52]. In another study, HA was modified by PEGylated dithiodipropionate dihydrazide (TPH) and loaded with DOX (DOX–PEG–SS–HA) to target a hepatic carcinoma cell. Here, DOX was released after stimulation by both reduction and pH conditions [53]. Xu *et al* synthesized ferrocene (FC)-conjugated chitosan oligosaccharide (COS). This FC–COS nanocarrier was sensitive to pH and oxidative conditions (NaClO or H_2O_2). Therefore, the loaded 5-fluorouracil (5FU) was released efficiently at pH 3.8 [54]. In another study, Zhao *et al* showed that targeted delivery of DOX can be achieved by reduction-sensitive GO nanocages. In the reductive milieu of endolysosomes (induced by high cysteine content) and in context of the acidic nature of endosomes and lysosomes (pH 4.5–6.5), the loaded DOX drugs were released [55].

Wen *et al* [56] synthesized a multi-controllable biopolymer nanogel nanocarrier that encapsulated hydrophobic anticancer drugs for intracellular targeting. They synthe-sized, via a facile aqueous free radical crosslinking polymerization, nanogels cross-linked with a disulfide linker (e.g. disulfide-labeled dimethacrylate (ssDMA)) of poly oligo(ethylene oxide)-containing methacrylate (POEOMA)-grafted carboxymethyl

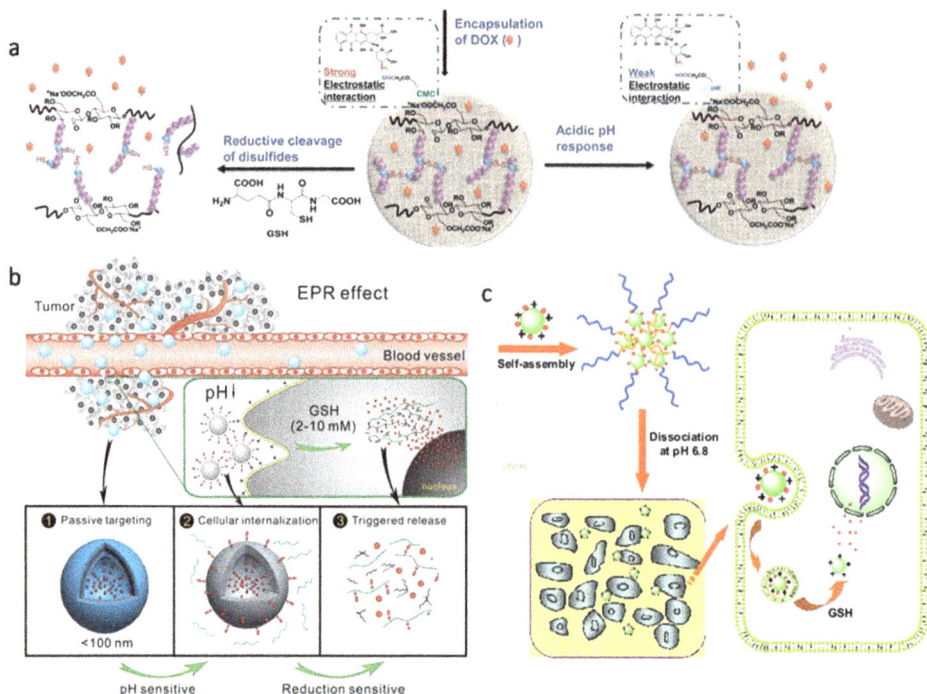

Figure 6.5. (a) Schematic of drug release in HeLa cancer cells from DOX-loaded biopolymer nanogel nanocarrier via cleavage of disulfide crosslinks in the reducing milieu of GSH and carboxylic acid groups in the acidic milieu of cancer cells. (b) Uptake by tumor cells of FPPLV containing reduction-responsive disulfide bonds and pH-responsive hydrazone bonds. (c) Response of polyplex micelles to mildly acidic and reducing milieu in a two-step drug delivery for deep tumor tissue penetration and the elimination of drug resistance to cisplatin. Reproduced with permission from [56, 57, 59], respectively. Copyright 2014 by The Royal Society of Chemistry.

cellulose (CMC). In this manner, the cleavage of disulfide crosslinks and the pH response of the carboxylic acid groups in the CMC could be obtained in reducing and acidic milieus, respectively. Thus, the DOX intracellular release inside HeLa cancer cells triggered by acidic pH and the presence of GSH was increased, with a more significant effect from low pH (figure 6.5(a)).

Wang *et al* [57, 58] studied the various dual stimuli pH-reduction-sensitive delivery systems for targeting tumor sites. For example, they synthesized 50 nm folate–PEG-coated polymeric lipid vehicles (FPPLV). In these, amphiphilic dextran derivatives were prepared in which stearyl alcohol (SA) chains with reduction-responsive disulfide bonds and PEG chains with pH-responsive hydrazone bonds were conjugated together. In acidic pH/reducing milieus, the hydrazone and disulfide bonds were cleaved, inducing a triggered DOX release. The *in vitro* cellular uptake study showed that through exposure of the folate in acidic conditions, the PEG coating was removed from FPPLV and the vehicles effectively internalized in the tumor cells via ligand receptor interactions. In addition, the nanocarriers were broken (triggered by intracellular GSH levels in tumor cells), resulting in selective drug release. Consequently, the FPPLV demonstrated significant antitumor activity against HeLa cells (figure 6.5(b)) [57].

Li *et al* [59] studied the delivery capability of multistage dual endogenous stimulus-sensitive polyplex micelles utilized for deep tumor tissue penetration and the elimination of the drug resistance of tumors against the cisplatin drug through a two-step anticancer DDS. The nanocarrier was synthesized via the self-assembly of anionic block copolymers and platinum (IV)-conjugated dendrimer prodrugs. In the acidic environment of tumor tissue (pH 6.8), charge conversion in pH-responsive anionic block copolymer occurs, leading to the fast dissociation of polyplex micelles as a result of electrostatic repulsion. Thereafter, in the reductive environment, the positively charged conjugated prodrugs can be released in the form of molecules with small size and high mobility, inducing deep penetration as well as appropriate dispersion in the tumor site because of their high affinity to the negative cellular membrane. Thus, the dendrimer prodrugs can be internalized into the tumor cells and can release the active cisplatin in the reducing cytosol (figure 6.5(c)). The effective penetration into dense tumor tissue and the enhanced cytotoxicity of the anticancer drug (compared to the insignificant toxicity of the dendrimers) contributed to remarkable growth inhibition in respect of A549R tumor cells.

6.2.5 pH/biomolecule-responsive

Combined pH/enzyme-triggered nanocarriers are one of the most studied pH/biomolecule dual stimuli-responsive DDS [15, 60]. Enzyme/pH-responsive NP were designed with a conjugation of Eudragit S100 (ES) and azo-PU to encapsulate coumarin-6 (C-6) for sustained release at physiological pH, while preventing rapid drug release in the small intestine and stomach [61]. The combination of HA and diethylaminopropyl (DEAP) was studied in order to produce dual-responsive micelles for the encapsulation of DOX. Herein, the protonation of DEAP in an acidic milieu occurred, followed by the degradation of HA by hyaluronidase activity in target cells, leading to the release of DOX. Drug release was reported to be maximal in acidic endosomes with high hyaluronidase activity [60]. A controlled insulin delivery vehicle was developed via the conjugation of glucose with concanavalin A (ConA) followed by binding to DMAEMA. This hydrogel displayed good sustained release of insulin, which was auto-regulated in the presence of glucose [62].

Acidic pH conditions and lysozyme enzymes are both inherently available in various targeted cancer cells, with lysozyme being produced extensively in them, especially in myelomonocytic leukemia. Hakeem *et al* [63] developed a nanovehicle based on a naturally degradable chitosan, a linear polysaccharide, covalently coupled with end-capped MCM-41 type MSN for anticancer drug delivery. It is known that a chitosan polymer is quickly hydrolyzed into its monomers in an acidic milieu and in the presence of lysozyme as a catalyst. Premature cargo release was controlled through the use of chitosan nanovalves as a gatekeeper to close the pores of the MSN in physiological conditions. Cargo release was triggered by exposure to lysozyme enzymes and an acidic environment.

Prasmod *et al* [64] designed pH/enzyme dual stimuli-responsive polysaccharide vesicular nanovehicles based on dextran for the delivery of DOX via both physically loaded and chemically conjugated forms within breast cancer cells. A modified dextran derivative was self-assembled into vesicular assemblies (~200 nm). The vesicular structure of the dextran backbone was formed by exploiting a 3-pentadecylphenol hydrophobic unit as a structure director. Meanwhile, the water-soluble anticancer drug doxorubicin (DOX·HCl) was encapsulated in the hydrophobic pocket of this nanovehicle and also anchored in the dextran backbone of the vesicular nanovehicles by imine moieties. Hence, the DOX concurrently loaded at the core and conjugated at the hydrophobic layer of the nascent polysaccharide vesicles (figure 6.6(a)). The imine chemical linkage and a lysosomal esterase enzyme-cleavable aliphatic ester bond connected the hydrophobic segment to the dextran and provided the pH- and enzyme-responsiveness of the nanovehicles. According to the results, the dextran-based nanovehicles showed suitable stability in circulatory physiological conditions (pH 7.4 and 37 °C). The acid-labile benzylic imine linkage was cleaved instantly at acidic pH ($\leqslant$6.0), while the esterase enzyme present in the lysosomal compartments of cells induced vesicular rupture. Therefore, 100% of the drug molecules (both conjugated and loaded) were released efficiently in the intracellular milieus (figure 6.6(b)). In addition, the cytotoxicity assay of the DOX·HCl-loaded dextran nanovehicles indicated non-cytotoxicity up to 500 μg mL^{-1}. The MTT tests on the fibroblast cells showed that cancer cell death was achieved efficiently using DOX-encapsulated (i.e. loaded and conjugated) nanovehicles compared to use of the free drug in MCF7 breast cancer cells.

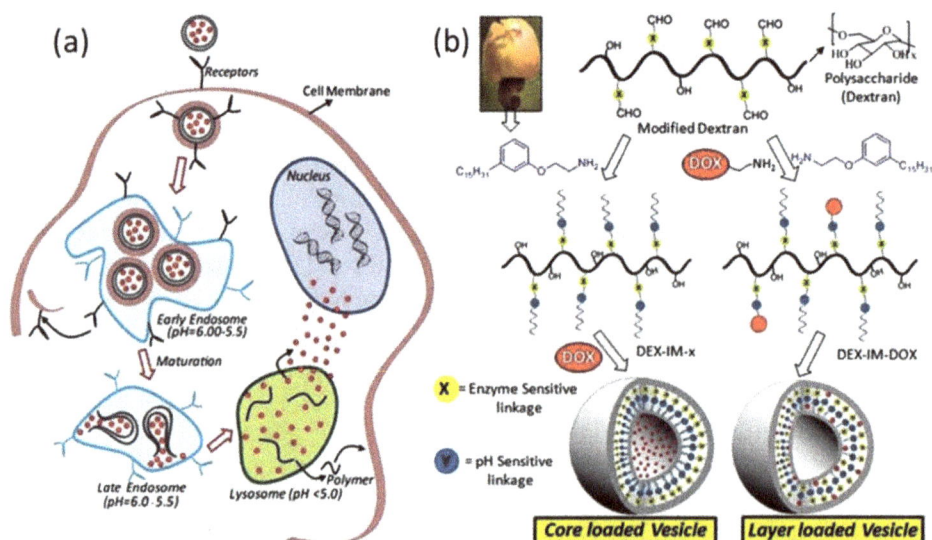

Figure 6.6. Schematic of (a) the structure of pH/enzyme dual stimuli-sensitive polysaccharide nanovehicles and (b) their cellular uptake and subsequent degradation in endocytic compartments. Reproduced with permission from [64]. Copyright 2015 by The Royal Society of Chemistry.

Glutathione is often employed in the design of pH/biomolecule dual-responsive DDS. Wang *et al* [65] studied glutathione/pH dual-sensitive single-hole degradable hollow silica NP (DHSN) with features that included high pore volume, facile functionalization, high colloidal stability and low density for the delivery of DOX to cancer cells. To impart glutathione sensitivity, the DHSN contained a disulfide bond-bridged silane (BTOCD), while to enhance intracellular delivery efficiency a pH-response element was also introduced. The results showed that DOX was loaded effectively into the DHSN. These nanocarriers can be degraded into pieces after exposure to glutathione. In addition, insignificant hemolytic activity and minor cytotoxicity indicated the high biocompatibility of the DHSN nanocarriers. Therefore, DOX was controllably released from the nanocarriers through their decomposition within the nuclei of TCA8113 cancer cells, triggered by the lower pH and the high concentration of glutathione.

The glucose biomolecule is often used to produce dual pH/biomolecule-responsive nanocarriers. Wu *et al* [66] synthesized a glucose/pH dual-sensitive Con A lectin protein-gated mannose carbohydrate-functionalized MSN nanocontainer. Carbohydrate–protein interactions play a key role in various cellular processes, such as cell adhesion, metastasis, inflammation, immune response and cell differentiation. Con A is a carbohydrate-binding lectin protein that can bind specifically to mannose and glucose epitopes in the presence of Ca^{2+} and Mg^{2+}. These carbohydrate-functionalized nanoparticle vehicles can provide protein recognition and sensing. The optimized functionalization of MSN with mannose ligands was carried out, followed by the construction of tight Con A tetramer nanogates bound to the functionalized mannose epitopes through multivalent carbohydrate–protein interactions, through which the cargo molecules were encapsulated within the pores. The local spatial arrangement of the mannose ligands was adjusted through the linked long and flexible spacers, leading to favorable multivalent protein binding. Thereafter, the cargo could be released on demand, triggered by exposure to an acidic environment (as found in tumor cells and inflammatory tissue) or through the competitive binding of glucose at an increased blood-glucose concentration (as seen in diabetes). The Con A protein nanogated carriers illustrated very low premature release of cargo.

6.3 Triple stimuli-based delivery systems

In recent years, there have been an increasing number of reports concerning multi-responsive NP that can respond to three or more integrated stimuli (internal and external). Here, we have gathered a few recent studies that employ internal stimuli as part of a triple stimuli-responsive DGDS for drug/gene targeting. Table 6.2 lists triple stimuli-responsive (drug/gene delivery) systems based on the integration of internal triggers with each other or with external stimuli.

In a study, Lee *et al* generated triple redox/photo/pH-responsive micelles employing pH-responsive boronate ester and dithiothreitol (DTT)-responsive disulfide bonds for triggered drug release, and a spiropyran-based photochromic polymer, (spiropyran/boronic acid-conjugated poly(dimethylamino ethyl methacrylate-co-methacrylic acid) (S-PMA)), as photo-responsive imaging agent for simultaneous cell-imaging. Here, phenylboronic acid-conjugated Pluronic was cross-linked with lactose-modified

Table 6.2. Triple stimuli-based smart systems for drug/gene delivery.

Stimuli	Nanocarrier	Synthesis method	Drug	Results/outcome	Reference
Thermo, pH and reduction	Micellar NP based on a block copolymer p (PEG-MEMA-co-Boc-Cyst-MMAm-co-VI)-b-PEG (PPBV) composed of hydrophobic units with disulfide linkages and removable tert-butyloxycarbonyl, hydrophilic units (PEGMEMA) and ionizable units (vinylimidazole)	Conventional radical polymerization	PTX	• The PPBV micelles showed high stability in PBS (pH 7.4, 37 °C) • PTX loaded into PPBV with high loading and encapsulation efficiency • Slow, sustained release in simulated extracellular conditions, immediate drug release in a weakly acidic or reductive environment • Non-toxicity of empty micelles and high cytotoxicity of PTX-loaded micelles against HepG2 cancer cells, as well as enhanced therapeutic efficacy	[67]
Thermo, redox and pH	Biocompatible poly (N-isopropylacrylamide-co-acrylic acid) nanogels	*In situ* polymerization of N-isopropylacrylamide (NIPAM) and acrylic acid (AA) in the presence of sodium dodecyl sulfate (SDS) as a surfactant, utilizing N,N′-bis (acryloyl)cystamine (BAC) (a biodegradable reducible crosslinker) or N,N′-methylenebisacrylamide (MBA) (a nondegradable crosslinker)	DOX	• DOX loaded in the nanogels via electrostatic interactions through simple mixing in aqueous solution • PNA–BAC nanogels showed higher DOX-loading capacity than nondegradable MBA nanogels • Drug release from PNA–BAC nanogels via thermal triggering; mimicked reductive intracellular conditions and acidic tumor microenvironment • Fast cellular uptake of nanogels by CAL-72 cells (an osteosarcoma cell line) followed by high DOX accumulation in cells and enhanced cytotoxicity compared to free DOX and DOX-loaded nondegradable nanogels	[68]

Table 6.2. (Continued.)

Stimuli	Nanocarrier	Synthesis method	Drug	Results/outcome	Reference
Glucose, pH and thermo	Reversibly crosslinked (RCL) nanogels made of thermo-responsive poly (vinyl alcohol)-b-poly (N-vinylcaprolactam) copolymers with boronate/diol bonding and surface-functionalized superparamagnetic maghemite NP	Hydrolysis of poly(vinyl acetate)-b-poly(N-vinyl-caprolactam) copolymers (PVAc-b-PNVCL), prepared by cobalt-mediated radical polymerization (CMRP) accompanied by a purification step	Hydrophobic drug model Nile red (NR)	• Zero premature release at physiological pH in the absence of glucose • Triggered release at acidic pH (5.0) and/or in the presence of glucose • Magnetically triggered drug release • Appropriate hydrophobic drug loading and encapsulation efficiency • Good biocompatibility for biomedical and theranostic applications • Enhanced inhibition of cell proliferation through pH/glucose triggered nanocarriers	[69]
Thermo, pH and light	Poly(N-isopropylacrylamide-co-methacrylic acid)-Au hybrid microgels	Soap-free emulsion polymerization and *in situ* reduction of gold precursor (i.e. gold–thiol chemistry) in the presence of thiol-functionalized poly (N-isopropylacrylamide-co-methacrylic acid) microgels	—	• Desirable swelling/deswelling transition of as-synthesized hydrogels triggered by temperature, pH and light irradiation • Modulation of the plasmonic properties of microgels through the stimuli-triggered volume phase transition • Tunable catalytic activity of the microgels	[70]

(*Continued.*)

Table 6.2. (Continued.)

Stimuli	Nanocarrier	Synthesis method	Drug	Results/outcome	Reference
Glucose, pH and redox	Assembled micelles with a PEO shell and a PAPBA core	Reversible addition fragmentation chain transfer polymerization (RAFT) synthesis of amphiphilic copolymer containing poly(ethylene oxide) (PEO), poly(3-acrylamidophenylboronic acid) (PAPBA) and a disulfide bond (PEO–SS–PAPBA)	Model drugs such as insulin	• pH alterations and glucose responses changed the hydrophobic core to hydrophilic after the addition of a base or glucose • Formation/dissociation of micelles through pH alteration • Dissociation of the micelles by cleavage of disulfide bonds corresponding to the thiols by reduction or disulfide–thiol exchange after the addition of GSH • Controlled release of insulin triggered by the addition of glucose or GSH • Glucose response near pK of PAPBA segments in the copolymer micelle	[71]

chitosan (Plu-Ch), and then S–PMA was conjugated to Plu–Ch to form Plu–Ch–S–PMA. This material was used for combined Taxol delivery/optical imaging in the cell environment under reductive conditions [72]. Qin *et al* [40] developed a light/temperature/pH triple stimuli-responsive DOX-loaded nanogel nanocarrier based on thermo-sensitive poly(N-isopropylacrylamide), CNT (as a NIR triggered motif) and amphiphilic alkyl chain-anchored chitosan (for the dispersion of CNT in aqueous solution) with a tumor targeting capability and minimal side-effects. These (CS/PNIPAAm@CNT) nanocarriers showed a high drug-loading capacity (43%) and rapid drug release occurred at 40 °C and pH 5.0 (compared to 25 °C and pH 7.4, respectively). Furthermore, NIR laser irradiation induced faster and repetitive DOX release from the nanocarrier. Enhanced cytotoxicity in the HeLa cells was obtained via the NIR-stimulated increase in temperature and the higher DOX release rate. Chang *et al* synthesized thermo-responsive PNIPAAm coupled with pH-sensitive Mac, and this was combined with magnetic NP to form a thermo/magnetic/pH tri-responsive carrier. These microspheres were utilized for the delivery of DOX into HeLa cells [73].

Liu *et al* [69] studied a pH, glucose and thermo triple-responsive DDS. This nanosystem was composed of nanogels crosslinked by functional superparamagnetic maghemite NP. Here, for the synthesis of reversibly crosslinked (RCL) nanogels, temperature-sensitive copolymer micelles of poly(vinylalcohol)-b-poly(N-vinylcaprolactam) (PVOH-b-PNVCL) were formed. Afterward, the maghemite NP (γ-Fe$_2$O$_3$ NP) were functionalized with 3-carboxy-5-nitro-phenylboronic acid (CNPBA) moieties and used to reversibly crosslink the PVOH by bonding the boronate/diol corona that was activated above the LCST of the PNVCL block. The hydrophobic drug molecules were loaded and trapped in the RCL nanogels at a pH of 7.4. The RCL nanogels showed minimal *in vivo* premature drug release at physiological pH in the absence of glucose. In addition, the activation of several stimuli through pH alterations (a decrease to an acidic pH of 5.0) and/or the presence of competing diols (e.g. glucose), or even through magnetic-responsive behavior from the presence of the superparamagnetic maghemite NP (inducing a magnetically generated heating), triggered the cleavage of boronate/diol bonding, leading to the release of drug molecules. Thereafter, cell proliferation in the DMEM complete medium was enhanced. Thus, these nanocarriers can also be used for MR imaging and remotely magnetically responsive accelerated drug release upon alternating magnetic field (AMF). Cytotoxicity against fibroblast-like L929 and human melanoma Mel-5 cell lines via the MTS assay illustrated the appropriate biocompatibility of the nanogels.

References

[1] Alvarez-Lorenzo C, Bromberg L and Concheiro A 2009 Light-sensitive intelligent drug delivery system *Photochem. Photobiol.* **85** 848–60
[2] Chan A, Orme R P, Fricker R A and Roach P 2013 Remote and local control of stimuli responsive materials for therapeutic applications *Adv. Drug Del. Rev.* **65** 497–514
[3] Wu L, Zou Y, Deng C, Cheng R, Meng F and Zhong Z 2013 Intracellular release of doxorubicin from core-crosslinked polypeptide micelles triggered by both pH and reduction conditions *Biomaterials* **34** 5262–72

[4] Cheng R *et al* 2011 Reduction and temperature dual-responsive crosslinked polymersomes for targeted intracellular protein delivery *J. Mater. Chem.* **21** 19013–20

[5] Kaushik A, Jayant R D, Sagar V and Nair M 2014 The potential of magneto-electric nanocarriers for drug delivery *Expert Opin. Drug Deliv.* **11** 1635–46

[6] Cheng R, Meng F, Deng C, Klok H-A and Zhong Z 2013 Dual and multi-stimuli responsive polymeric nanoparticles for programmed site-specific drug delivery *Biomaterials* **34** 3647–57

[7] Hu J, Meng H, Li G and Ibekwe S I 2012 A review of stimuli-responsive polymers for smart textile applications *Smart Mater. Struct.* **21** 053001

[8] Yang Z *et al* 2015 Drug and gene co-delivery systems for cancer treatment *Biomaterials Science* **3** 1035–49

[9] Shao Y, Shi C, Xu G, Guo D and Luo J 2014 Photo and redox dual responsive reversibly cross-linked nanocarrier for efficient tumor-targeted drug delivery *ACS Appl. Mater. Interf.* **6** 10381–92

[10] Cheng X, Jin Y, Sun T, Qi R, Fan B and Li H 2015 Oxidation- and thermo-responsive poly (N-isopropylacrylamide-co-2-hydroxyethyl acrylate) hydrogels cross-linked via diselenides for controlled drug delivery *RSC Adv.* **5** 4162–70

[11] Zhou T, Zhao X, Liu L and Liu P 2015 Preparation of biodegradable PEGylated pH/reduction dual-stimuli responsive nanohydrogels for controlled release of an anti-cancer drug *Nanoscale* **7** 12051–60

[12] Shi X, Zheng Y, Wang G, Lin Q and Fan J 2014 pH- and electro-response characteristics of bacterial cellulose nanofiber/sodium alginate hybrid hydrogels for dual controlled drug delivery *RSC Adv.* **4** 47056–65

[13] Sharker S M, Kim S M, Kim S H, In I, Lee H and Park S Y 2015 Target delivery of β-cyclodextrin/paclitaxel complexed fluorescent carbon nanoparticles: externally NIR light and internally pH sensitive-mediated release of paclitaxel with bio-imaging *J. Mater. Chem.* B **3** 5833–41

[14] Lee C-S and Na K 2014 Photochemically triggered cytosolic drug delivery using pH-responsive hyaluronic acid nanoparticles for light-induced cancer therapy *Biomacromolecules* **15** 4228–38

[15] Chen X, Soeriyadi A H, Lu X, Sagnella S M, Kavallaris M and Gooding J J 2014 Dual bioresponsive mesoporous silica nanocarrier as an "AND" logic gate for targeted drug delivery to cancer cells *Adv. Funct. Mater.* **24** 6999–7006

[16] Salehi R, Hamishehkar H, Eskandani M, Mahkam M and Davaran S 2014 Development of dual responsive nanocomposite for simultaneous delivery of anticancer drugs *J. Drug Target.* **22** 327–42

[17] Yuan H, Li B, Liang K, Lou X and Zhang Y 2014 Regulating drug release from pH- and temperature-responsive electrospun CTS-g-PNIPAAm/poly (ethylene oxide) hydrogel nano-fibers *Biomed. Mater.* **9** 055001

[18] Singh N K and Lee D S 2014 *In situ* gelling pH- and temperature-sensitive biodegradable block copolymer hydrogels for drug delivery *J. Controlled Release* **193** 214–27

[19] Yang C *et al* 2014 pH-responsive magnetic core–shell nanocomposites for drug delivery *Langmuir* **30** 9819–27

[20] Yang L, Shi J, Zhou X and Cao S 2015 Hierarchically organization of biomineralized alginate beads for dual stimuli-responsive drug delivery *Int. J. Biol. Macromol.* **73** 1–8

[21] Yin H and Casey P S 2014 Effects of iron or manganese doping of ZnO nanoparticles on their dissolution, ROS generation and cytotoxicity *RSC Adv.* **4** 26149–57

[22] Efthimiadou E K, Tapeinos C, Tziveleka L-A, Boukos N and Kordas G 2014 pH- and thermo-responsive microcontainers as potential drug delivery systems: morphological characteristic, release and cytotoxicity studies *Mater. Sci. Eng.* C **37** 271–7

[23] Chiang W-L *et al* 2015 A rapid drug release system with a NIR light-activated molecular switch for dual-modality photothermal/antibiotic treatments of subcutaneous abscesses *J. Controlled Release* **199** 53–62

[24] Rodkate N, Rutnakornpituk B, Wichai U, Ross G and Rutnakornpituk M 2015 Smart carboxymethylchitosan hydrogels that have thermo-and pH-responsive properties *J. Appl. Polym. Sci.* **132** 8

[25] Mu B and Liu P 2012 Temperature and pH dual responsive crosslinked polymeric nanocapsules via surface-initiated atom transfer radical polymerization *React. Funct. Polym.* **72** 983–9

[26] Joshi R V, Nelson C E, Poole K M, Skala M C and Duvall C L 2013 Dual pH- and temperature-responsive microparticles for protein delivery to ischemic tissues *Acta Biomater.* **9** 6526–34

[27] Huynh C T, Nguyen Q V, Kang S W and Lee D S 2012 Synthesis and characterization of poly (amino urea urethane)-based block copolymer and its potential application as injectable pH/temperature-sensitive hydrogel for protein carrier *Polymer* **53** 4069–75

[28] Sundaresan V, Menon J U, Rahimi M, Nguyen K T and Wadajkar A S 2014 Dual-responsive polymer-coated iron oxide nanoparticles for drug delivery and imaging applications *Int. J. Pharmaceut.* **466** 1–7

[29] Patra S, Roy E, Karfa P, Kumar S, Madhuri R and Sharma P K 2015 Dual-responsive polymer coated superparamagnetic nanoparticle for targeted drug delivery and hyperthermia treatment *ACS Appl. Mater. Interf.* **7** 9235–46

[30] Li X *et al* 2014 A novel injectable pH/temperature sensitive CS-HLC/β-GP hydrogel: the gelation mechanism and its properties *Soft Mater.* **12** 1–11

[31] Weaver C L, LaRosa J M, Luo X and Cui X T 2014 Electrically controlled drug delivery from graphene oxide nanocomposite films *ACS Nano* **8** 1834–43

[32] Liu X, Liu H-J, Cheng F and Chen Y 2014 Preparation and characterization of multi stimuli-responsive photoluminescent nanocomposites of graphene quantum dots with hyper-branched polyethylenimine derivatives *Nanoscale* **6** 7453–60

[33] Chen Y *et al* 2014 Multifunctional graphene oxide-based triple stimuli-responsive nano-theranostics *Adv. Funct. Mater.* **24** 4386–96

[34] Yang X *et al* 2011 Multi-functionalized graphene oxide based anticancer drug-carrier with dual-targeting function and pH-sensitivity *J. Mater. Chem.* **21** 3448–54

[35] Sun S and Wu P 2011 A one-step strategy for thermal- and pH-responsive graphene oxide interpenetrating polymer hydrogel networks *J. Mater. Chem.* **21** 4095–7

[36] Shim W S, Yoo J S, Bae Y H and Lee D S 2005 Novel injectable pH and temperature sensitive block copolymer hydrogel *Biomacromolecules* **6** 2930–4

[37] Singh N K and Lee D S 2014 *In situ* gelling pH- and temperature-sensitive biodegradable block copolymer hydrogels for drug delivery *J. Controlled Release* **193** 214–27

[38] Huynh C T, Nguyen M K and Lee D S 2011 Biodegradable star-shaped poly (ethylene glycol)-poly (β-amino ester) cationic pH/temperature-sensitive copolymer hydrogels *Colloid Polym. Sci.* **289** 301–8

[39] Huynh C T, Nguyen M K and Lee D S 2011 Biodegradable pH/temperature-sensitive oligo (β-amino ester urethane) hydrogels for controlled release of doxorubicin *Acta Biomater.* **7** 3123–30

[40] Qin Y *et al* 2015 Near-infrared light remote-controlled intracellular anti-cancer drug delivery using thermo/pH sensitive nanovehicle *Acta Biomater.* **17** 201–9

[41] Knezevic N Z, Trewyn B G and Lin V S 2011 Light- and pH-responsive release of doxorubicin from a mesoporous silica-based nanocarrier *Chemistry* **17** 3338–42

[42] Feng N, Han G, Dong J, Wu H, Zheng Y and Wang G 2014 Nanoparticle assembly of a photo- and pH-responsive random azobenzene copolymer *J. Colloid Interface Sci.* **421** 15–21

[43] Yu J-C *et al* 2014 pH-Responsive and near-infrared-emissive polymer nanoparticles for simultaneous delivery, release, and fluorescence tracking of doxorubicin *in vivo Chem. Commun.* **50** 4699–702

[44] Yadavalli T, Ramasamy S, Chandrasekaran G, Michael I, Therese H A and Chennakesavulu R 2014 Dual responsive PNIPAM–chitosan targeted magnetic nanopolymers for targeted drug delivery *J. Magn. Magn. Mater.* **380** 315–20

[45] Yu S *et al* 2013 Magnetic and pH-sensitive nanoparticles for antitumor drug delivery *Colloid. Surface.* B **103** 15–22

[46] Kang X J *et al* 2012 Poly (acrylic acid)-modified Fe_3O_4 microspheres for magnetic-targeted and ph-triggered anticancer drug delivery *Chemistry* **18** 15676–82

[47] Mu B, Liu P, Du P, Dong Y and Lu C 2011 Magnetic-targeted pH-responsive drug delivery system via layer-by-layer self-assembly of polyelectrolytes onto drug-containing emulsion droplets and its controlled release *J. Polym. Sci. Pol. Chem.* **49** 1969–76

[48] Bui Q N, Li Y, Jang M-S, Huynh D P, Lee J H and Lee D S 2015 Redox- and pH-sensitive polymeric micelles based on poly (β-amino ester)-grafted disulfide methylene oxide poly (ethylene glycol) for anticancer drug delivery *Macromolecules* **48** 4046–54

[49] Sun H, Meng F, Cheng R, Deng C and Zhong Z 2014 Reduction and pH dual-bioresponsive crosslinked polymersomes for efficient intracellular delivery of proteins and potent induction of cancer cell apoptosis *Acta Biomater.* **10** 2159–68

[50] Bai L, Wang X-h, Song F, Wang X-l and Wang Y-z 2014 'AND' logic gate regulated pH and reduction dual-responsive prodrug nanoparticles for efficient intracellular anticancer drug delivery *Chem. Commun.* **51** 93–6

[51] Li D, Zhang Y, Jin S, Guo J, Gao H and Wang C 2014 Development of a redox/pH dual stimuli-responsive MSP@ P (MAA-Cy) drug delivery system for programmed release of anticancer drugs in tumour cells *J. Mater. Chem.* B **2** 5187–94

[52] Gao C *et al* 2014 pH/redox responsive core cross-linked nanoparticles from thiolated carboxymethyl chitosan for *in vitro* release study of methotrexate *Carbohyd. Polym.* **111** 964–70

[53] Xu M *et al* 2013 Reduction/pH dual-sensitive PEGylated hyaluronan nanoparticles for targeted doxorubicin delivery *Carbohyd. Polym.* **98** 181–8

[54] Wu H *et al* 2014 Prostate stem cell antigen antibody-conjugated multiwalled carbon nanotubes for targeted ultrasound imaging and drug delivery *Biomaterials* **35** 5369–80

[55] Zhao X, Liu L, Li X, Zeng J, Jia X and Liu P 2014 Biocompatible graphene oxide nanoparticle-based drug delivery platform for tumor microenvironment-responsive triggered release of doxorubicin *Langmuir* **30** 10419–29

[56] Wen Y and Oh J K 2014 Dual-stimuli reduction and acidic pH-responsive bionanogels: intracellular delivery nanocarriers with enhanced release *RSC Adv.* **4** 229–37

[57] Wang S *et al* 2014 Smart pH- and reduction-dual-responsive folate–PEG-coated polymeric lipid vesicles for tumor-triggered targeted drug delivery *Nanoscale* **6** 7635–42

[58] Wang S *et al* 2014 pH- and reduction-responsive polymeric lipid vesicles for enhanced tumor cellular internalization and triggered drug release *ACS Appl. Mater. Interf.* **6** 10706–13

[59] Li J *et al* 2014 Dual endogenous stimuli-responsive polyplex micelles as smart two-step delivery nanocarriers for deep tumor tissue penetration and combating drug resistance of cisplatin *J. Mater. Chem.* B **2** 1813–24

[60] Kim S W, Oh K T, Youn Y S and Lee E S 2014 Hyaluronated nanoparticles with pH- and enzyme-responsive drug release properties *Colloid. Surface.* B **116** 359–64

[61] Naeem M, Kim W, Cao J, Jung Y and Yoo J-W 2014 Enzyme/pH dual sensitive polymeric nanoparticles for targeted drug delivery to the inflamed colon *Colloid. Surface.* B **123** 271–8

[62] Yin R, Tong Z, Yang D and Nie J 2011 Glucose and pH dual-responsive concanavalin A based microhydrogels for insulin delivery *Int. J. Biol. Macromol.* **49** 1137–42

[63] Hakeem A *et al* 2014 Dual stimuli-responsive nano-vehicles for controlled drug delivery: mesoporous silica nanoparticles end-capped with natural chitosan *Chem. Commun.* **50** 13268–71

[64] Pramod P, Shah R and Jayakannan M 2015 Dual stimuli polysaccharide nanovesicles for conjugated and physically loaded doxorubicin delivery in breast cancer cells *Nanoscale* **7** 6636–52

[65] Wang D *et al* 2014 Fabrication of single-hole glutathione-responsive degradable hollow silica nanoparticles for drug delivery *ACS Appl. Mater. Interf.* **6** 12600–8

[66] Wu S, Huang X and Du X 2013 Glucose- and pH-responsive controlled release of cargo from protein-gated carbohydrate-functionalized mesoporous silica nanocontainers *Angew. Chem.* **125** 5690–4

[67] Huang X *et al* 2013 Triple-stimuli (pH/thermo/reduction) sensitive copolymers for intracellular drug delivery *J. Mater. Chem.* B **1** 1860–8

[68] Zhan Y *et al* 2015 Thermo/redox/pH-triple sensitive poly (N-isopropylacrylamide-co-acrylic acid) nanogels for anticancer drug delivery *J. Mater. Chem.* B **3** 4221–30

[69] Liu J *et al* 2014 Glucose-, pH- and thermo-responsive nanogels crosslinked by functional superparamagnetic maghemite nanoparticles as innovative drug delivery systems *J. Mater. Chem.* B **2** 1009–23

[70] Shi S, Wang Q, Wang T, Ren S, Gao Y and Wang N 2014 Thermo-, pH-, and light-responsive poly (N-isopropylacrylamide-co-methacrylic acid)–Au hybrid microgels prepared by the *in situ* reduction method based on Au–thiol chemistry *J. Phys. Chem.* B **118** 7177–86

[71] Yuan W, Shen T, Wang J and Zou H 2014 Formation–dissociation of glucose, pH and redox triply responsive micelles and controlled release of insulin *Polymer Chem.* **5** 3968–71

[72] Lee S Y, Lee H, In I and Park S Y 2014 pH/redox/photo responsive polymeric micelle via boronate ester and disulfide bonds with spiropyran-based photochromic polymer for cell imaging and anticancer drug delivery *Eur. Polym. J.* **57** 1–10

[73] Chang B, Sha X, Guo J, Jiao Y, Wang C and Yang W 2011 Thermo and pH dual responsive, polymer shell coated, magnetic mesoporous silica nanoparticles for controlled drug release *J. Mater. Chem.* **21** 9239–47

Chapter 7

Future perspectives and the global drug delivery systems market

In the last few decades, advances in nanomedicine, biotechnology, chemistry and materials science have re-ignited scientific and industrial interest in drug delivery research in an exponential manner. Although in the previous chapters, a great number of strategies for site-specific drug delivery and release have been reviewed, many investigators in the field of drug delivery face a dilemma between focusing on making an ever wider variety of DDS, or alternatively concentrating on enhancing the level of knowledge of the drug delivery process as a whole to eventually advance the use of the materials we already have to clinical applications in patients. In this regard, creating a multifunctional micro/nanocarrier system and preventing side-effects should be the mission of everyone concerned. In order for efficient drug/gene carriers to advance to the clinic, future studies should meet the following criteria:

- the release of cargo should be adjusted to the required condition to minimizing the side-effects;
- internal stimuli-responsive carriers should be regulated by such factors as design, morphology, chemical composition and carrier size in order to optimize the performance of developed systems;
- the synthesized DDS must not produce any reactions with other organs;
- the general toxicity to the environment and humans should be studied and, if necessary, removed.

Despite the numerous achievements that have been reported in recent years, it is crystal clear that these criteria are often closely interlinked and can be very challenging. It is generally accepted that these challenges will be solved by a multidisciplinary combination of different fields, such as materials characterization,

doi:10.1088/978-1-6817-4257-1ch7
7-1

physico-chemical analysis, nanotechnology and surface engineering. As far as the future scope of internal stimuli-responsive DDS is concerned, there are important aspects that still remain to be focused on:

- modification of the physico-chemical properties of the materials;
- improvement of DDS performance through computational modeling of nanoparticle targeted drug delivery;
- functionalization of the carrier surface;
- development of targeted internal-responsive DDS that provide more localized drug/gene delivery;
- minimization of damage to non-target tissues by enhancement of the therapeutic drug action of carriers and their biological responses;
- synthesis of novel materials as biodegradable polymers in order to achieve efficient delivery of insulin.

Recent achievements in DDS in clinical applications/the late experimental stages are another interesting field for future studies. For instance, recent developments in polymer technology, chemistry and mechanical engineering have introduced many invaluable implantable DDS for a wide range of clinical applications (e.g. diabetic macular edema, drug eluting, retinitis pigmentosa (RP), AMD and retinal vein occlusion).

Researchers who work on clinical applications have reported that some drugs or materials can prevent the adhesion of microorganisms by eliminating the formation of biofilms. They have also regulated the response of tissues to devices by controlling the reactions against a 'foreign body'. As far as future scope is concerned, these clinical devices may release a drug to tissues in an efficient and controlled fashion that would hardly be achievable otherwise. The authors strongly suggest that future studies should focus on how this integration into a single system of drugs and medical devices can provide synergistic outcomes.

After what has been mentioned in the previous chapters, it is interesting to consider the current and future state of the smart DDS market. Undoubtedly, DDS represent one of the most interesting fields for investment, so that in recent years a great number of companies have focused on the aforementioned challenges in order to produce efficient DDS.

Such companies advocate the advantages of advanced DDS over traditional drug formulations, and concentrate on the synthesis of more effective drug and gene delivery systems for different applications, such as diabetes, rectal administration, lymphatic implantation, and also vaginal and oral delivery. Increasing the duration of drug activity and its efficiency, minimizing the side-effects by decreasing the toxicity and dosing frequency of drugs, and improving the targeting ability of the systems, are the main goals of these medical companies.

Reviewing the published reports on the global DDS market shows that research-based pharmaceutical companies are increasingly investing in DDS. For instance, based on the recent market report of BCC Research[1], the global DDS market

---

[1] www.bccresearch.com.

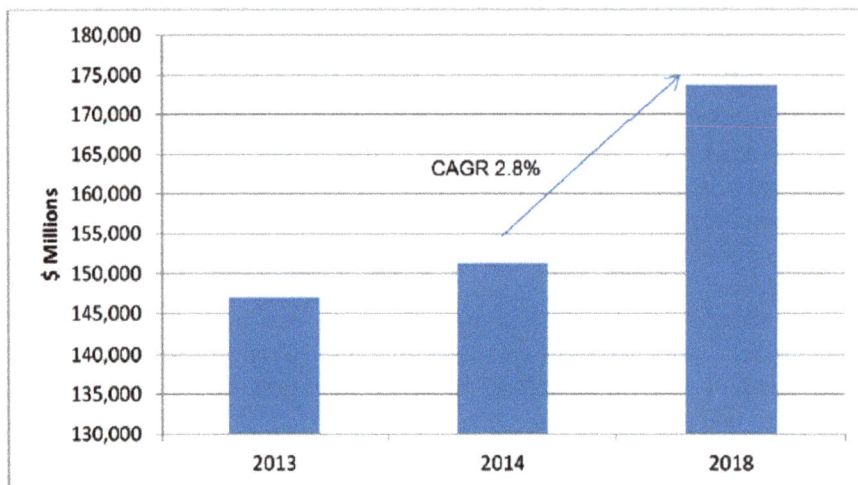

Figure 7.1. Current and predicted figures for advanced DDS global revenue. Source: BCC Research (PHM006J).

was valued at $151.3 billion US dollars in 2013. However, the five-year compound annual growth rate (CAGR) of this emerging market is calculated to be 2.8%. So, this report predicted that the market for DDS will reach $173.8 billion US dollars by 2018. Figure 7.1 presents the global revenue of advanced DDS. The development of new materials and technologies, the alliances between the private sector and universities, and increasing competition are the main factors driving this growth.

In addition, a similar trend has been reported in the field of gene delivery. According to the BCC Research report, although many challenges exist for the full-scale clinical development of nanoparticle-based RNAi delivery systems, positive growth has been seen in the market for GDS. For example, Dicerna, a pioneer company in RNAi therapeutics, has focused on Dicer substrate-based RNAi therapeutics since 2014. Although some companies like Novartis have decided to leave this market because of an unclear market outlook and the lack of suitable delivery technologies, BCC Research forecast that the global market for RNAi therapeutics will reach around $38.8 billion US dollars by 2018, thanks to the investments of such pioneering companies as Tekmira Pharmaceuticals, Quark Pharmaceuticals, Dicerna Pharmaceuticals, Silence Therapeutics and Alnylam Pharmaceuticals.

www.ingramcontent.com/pod-product-compliance
Lightning Source LLC
Chambersburg PA
CBHW081549220326
41598CB00036B/6614